T0320959

Sequence Spaces and Summability over Valued Fields

Sequence Spaces and Summability over Valued Fields

P.N. Natarajan

CRC Press
Taylor & Francis Group
Boca Raton London New York

CRC Press is an imprint of the
Taylor & Francis Group, an **informa** business

A CHAPMAN & HALL BOOK

CRC Press
Taylor & Francis Group
6000 Broken Sound Parkway NW, Suite 300
Boca Raton, FL 33487-2742

International Standard Book Number-13: 978-0-3672-3662-5 (Hardback)

Library of Congress Cataloging-in-Publication Data

Names: Natarajan, P. N., author.
Title: Sequence spaces and summability over valued fields / P.N. Natarajan.
Description: Boca Raton : CRC Press, Taylor & Francis Group, 2020. |
Includes bibliographical references and index.
Identifiers: LCCN 2019012736| ISBN 9780367236625 (hardback : alk.
paper) | ISBN 9780429281105 (ebook)
Subjects: LCSH: Linear topological spaces. | Sequence spaces. | Topological fields. | Valued fields.
Classification: LCC QA322 .N3865 2020 | DDC 515/.73--dc23
LC record available at https://lccn.loc.gov/2019012736

Visit the Taylor & Francis Web site at
http://www.taylorandfrancis.com

and the CRC Press Web site at
http://www.crcpress.com

Dedicated

to my mentor

Prof. Dr. M.S. Rangachari

Contents

About the Author

P.N. Natarajan is former Professor and Head, Department of Mathematics, Ramakrishna Mission Vivekananda College, Chennai. He received his Ph.D. from the University of Madras, under Prof. M.S. Rangachari, Former Director and Head, The Ramanujan Institute for Advanced Study in Mathematics, University of Madras. An active researcher, Prof. Natarajan has more than 100 research papers to his credit published in international journals like *Proceedings of the American Mathematical Society, Bulletin of the London Mathematical Society, Indagationes Mathematicae, Annales Mathematiques Blaise Pascal and Commentationes Mathematicae.* He has so far written 4 books and contributed a chapter each to two edited volumes, published by the famous international publishers Springer, Taylor & Francis, and Wiley. His research interests include Summability Theory and Functional Analysis, both classical and ultrametric. Prof. Natarajan was honoured with the Dr. Radhakrishnan Award for the Best Teacher in Mathematics for the 1990-1991 by the Government of Tamil Nadu. Besides visiting several institutes of repute in Canada, France, Holland and Greece on invitation, Prof. Natarajan has participated in several international conferences and has chaired sessions.

Foreword

Three quotations from Francis Bacon are frequently emphasized by scholars. All the three quotations mentioned here occur in his essay entitled "Of Studies." The first one is related to the classification of books in general. It is as follows:

"Read not to contradict and confute; nor to believe and take for granted; nor to find talk and discourse; but to weigh and consider."

The second one in the same essay is as follows:

"Some books are to be tasted, others to be swallowed, and some few to be chewed and digested; that is, some books are to be read only in parts; others to be read, but not curiously; and some few to be read wholly, and with diligence and attention."

The third shows the purpose of reading. It is to be taken very seriously. It is as follows:

"Reading maketh a full man; conference a ready man; and writing an exact man."

The book by Professor P.N. Natarajan has the title *Sequence Spaces and Summabaility over Valued Fields*. Mathematics scholars, who would be interested in the book fall into two categories. For some of these scholars in the first category, getting to know the topics covered in this book are more important than a requirement for intensive study with an intention of working and contributing to that field. They will be guided by the principle, "some books are to be read only in parts." The second category of scholars will be guided by the principle of making contributions to the field based on the work of the author P.N. Natarajan. They will want to "chew and digest the book," and hence would read the above-mentioned book "wholly with diligence and attention."

If these two categories of scholars are to be served, what is needed is an adequate admirable preface. The preface is an excellent part of this book. It provides a detailed account in the form of a short survey of the historical development of the subject and makes it clear that within the same book, matrix transformations between involved spaces are discussed under Archimedean and non-Archimedean valued fields. A reader who wants to know the major differences in results and techniques that are needed to prove those results should thoroughly work on all the chapters found in this book. This preface also discusses specific topics like the "Schur matrix," "Steinhaus theorems," and "Sliding hump technique." If the preface is to be used by a reader as to whether the book is going to be important for that reader, the provided preface does exactly that job. The preface ends with an adequate bibliography on the subject.

The book contains eight chapters. The material contained in each chapter is exactly along the lines of the short account provided in the preface for that chapter. Further a detailed bibliography at the end of each chapter (except Chapter 1) under the heading "References" will help serious-minded scholars from the second category to seek out the necessary sources.

Reviewing the contents of each chapter is totally unnecessary as the task would merely be rehashing those contents in a different way. The first category scholars can grasp the necessary needed concepts and quickly absorb the material presented in the book based on their inclinations. The serious-minded scholars that belong to the second category will be amply rewarded for their diligence and efforts later on if they choose to make contributions to the field.

Some mathematicians may be inclined to think that non-Archimedean matrix transformations are merely investigations directed to finding analogues of results found in the classical cases of Archimedean-valued fields. But these mathematicians conveniently forget the development of non-Euclidean

Geometry in the plane. The parallel axiom is the point of departure between these two subjects. If non-Euclidean Geometry requires mathematical investigation, the study of matrix transformations between non-Archimedean topological vector spaces also should deserve the same considerations.

As one of his predecessors, I have to nominate P.N. Natarajan as a research mathematician who unified, modified, and changed the course of research work in this field.

Having written a few other books before, the author has perfected his craft, thereby becoming "an exact man." Francis Bacon is right.

A book of this nature that covers Functional Analysis topics both conventional and the non-conventional makes the book highly desirable and useful to the mathematical community. Nearly twenty years ago the author P.N. Natarajan provided me a copy of his Ph.D. thesis. Due to frequent use, the pages of that copy are so brittle that I keep them in a sufficiently big Ziploc® bag. At that time I used to wish for a book of this nature. Sometimes it is gratifying to note that my wishful thinking may really materialize.

The author has rightly dedicated the book to Prof. Dr. M.S. Rangachari, who was his Ph.D. thesis supervisor. It is a book that should be available to students, specializing in Mathematics, in their college and university libraries.

March 3, 2019

V.K. Srinivasan
Professor Emeritus
University of Texas at El Paso
El Paso, Texas 79968-0514, U.S.A.
Residential address: V.K. Srinivasan
2915, Mary St., La Crescenta, CA.,
91214, U.S.A.

Preface

Study of infinite matrix transformations in the field $\mathbb{R}$ of real numbers or the field $\mathbb{C}$ of complex numbers is quite an old one. Numerous authors have studied general matrix transformations or matrix transformations pertaining to certain special classes for the past several decades. On the other hand, the study of matrix transformations in non-Archimedean valued fields is of a comparatively recent origin. In spite of the pioneering work of A.F. Monna relating to Analysis and Functional Analysis over non-Archimedean valued fields dating back to 1940s, it was only in 1956 that Andree and Petersen [1] proved the analogue of the Silverman-Toeplitz theorem for the p-adic field $\mathbb{Q}_p$, for a prime p, on the regularity of an infinite matrix transformation. Even thereafter, study of matrix transformations in non-Archimedean valued fields was at a slow pace. Roberts [15] proved the Silverman-Toeplitz theorem for a general non-Archimedean valued field, while Monna ([9, p. 127]) proved the same theorem using the analogue of the Uniform Boundedness Principle. After Monna, we have the papers of Rangachari and Srinivasan [14] and Srinivasan [18], and recently Somasundaram [17] took up the study of matrix transformations in non-Archimedean valued fields. Since the 1970s, Natarajan has been working extensively on sequence spaces, matrix transformations and special summability methods in non-Archimedean valued fields. Natarajan [12] proved the Silverman-Toeplitz theorem without using functional analytic tools, using a technique due to Schur [16], later fortified by Ganapathy Iyer [6]. This technique, which is now known as the "sliding hump method," was used earlier in [10]. In this context, it is worth pointing out that the absence of an analogue in non-Archimedean analysis for the signum function in classical analysis possibly made Roberts and Monna use functional analytic tools to prove the Silverman-Toeplitz theorem in non-Archimedean valued fields.

The literature so far available does not present a systematic study of the topic under consideration pointing out the analogues of the classical study as well as the deviations therefrom. In a private communication, Monna had frankly expressed that one reason for the laxity of such a study was perhaps the fact that some of the results in non-Archimedean analysis are very much analogous or simpler to prove than their classical counterparts. However, it is not true that every result in non-Archimedean analysis has a proof analogous to its classical counterpart or even a simpler proof. As pointed out earlier, the absence of an analogue for the signum function, forces us to search for alternate devices. These devices provide an entirely different proof of even an exact analogue of a classical theorem, as for instance, in the case of the famous Silverman-Toeplitz theorem [12] and Schur's theorem [10].

The main object of the present book is two-fold. First, we supplement substantially the comparatively small bulk of literature relating to sequence spaces and matrix transformations in non-Archimedean analysis. Secondly, we illustrate how new proofs are necessary for proving the exact analogues of classical results. Deviations from the classical case are also pointed out with illustrations.

Throughout the book, depending on the relevance, we treat a non-Archimedean field along with a general valued field except when results relating to $\mathbb{R}$ (the field of real numbers) or $\mathbb{C}$ (the field of complex numbers) are too familiar to be stated. A common fiber going through most of the chapters is the Ganapathy Iyer-Schur technique or the sliding hump method mentioned at the outset. A further streamlining of this technique is indicated in the proof of Theorem 38. When $K = \mathbb{R}$ or $\mathbb{C}$, the present book gives many results which are new.

The motivation for writing the present book is [11]. The research work of the author, scattered in different publications, have been brought together in the form of the present book. This material has been modified, wherever possible, to suit the style and theme of the current book.

The contents of the chapters of the book are now summarized.

Chapter 1 is aimed at making the book self-contained to the extent possible. Besides recalling familiar concepts and results, it provides a brief introduction to non-Archimedean analysis.

Chapter 2 records the truth of the Steinhaus theorem [10] in the non-Archimedean case too. In view of an earlier incorrect assertion ([17], p. 171) to the contrary, this record seems to be essential. For the sake of generality, certain sequence spaces, called in the sequel Λ_r, containing the space of Cauchy sequences are considered. The space Λ_r precisely contains sequences of 0s and 1s which are periodic with period r after a stage. In view of the Steinhaus theorem which is, in fact, true in the form that there exists a sequence of 0s and 1s not summable by a regular matrix, the summability or otherwise of the sequences of 0s and 1s in Λ_r is of interest. More explicitly it is pointed out that $A \in (\ell_\infty, c)$, i.e., A is a Schur matrix if and only if A sums all non-periodic sequences of 0s and 1s. For the sake of completeness, the classical analogue of this result is treated. Toward the conclusion of Chapter 2, a Steinhaus-type theorem, improving the Steinhaus theorem, is proved in the classical case. It is also pointed out that the analogue of this result fails to hold in the non-Archimedean case.

The main result in Chapter 3 is the characterization of the matrix class $(\ell_\alpha, \ell_\alpha)$, $\alpha > 0$ in the non-Archimedean case. Because of the fact that there is, as such, no classical analogue for this result, this result seems to be important. The proof of this result does not, however, give a clue as to how its classical counterpart could be proved. In the context of a non-Archimedean valued field, the Cauchy product series of any two convergent series is again convergent; also such a product of two sequences in ℓ_α, $\alpha > 0$, is again in ℓ_α. The latter fact fails to hold for $\alpha > 1$, when the field is Archimedean. In this chapter, we also prove a Mercerian theorem by considering the structure of the space $(\ell_\alpha, \ell_\alpha)$, $\alpha \geq 1$, of infinite matrices, when the field is non-Archimedean. Here again there is a discordant note about the failure of a Steinhaus-type theorem analogous to the one proved by Fridy [4] in the classical case. However, a suitable modification is made to retain the Steinhaus-type theorem in a general form.

In Chapter 4, we treat the analogues of classical results in the nature of characterization of regular and Schur matrices as also the characterization of convergent sequences in terms of the behavior of subsequences or rearrangements. Buck [2] proved that a complex sequence is convergent if and only if there exists a regular matrix which sums all of its subsequences. Maddox [8] showed that a complex matrix is a Schur matrix if and only if it sums all the subsequences of a bounded, divergent sequence. It is shown that both these results continue to hold in the non-Archimedean situation too. A result of Fridy [5], where "subsequence" is replaced by "rearrangement" in Buck's result is also shown to hold for the non-Archimedean case. However, as regards sequences in ℓ_1, it is shown that analogues of the results for the complex case established by Fridy [5] and Keagy [7] fail to hold in the non-Archimedean case. An attempt is made to salvage the analogue of Fridy's result in a general form. We characterize regular matrices by means of the core or kernel of sequences. In the case of complex sequences, the containment of the core of the transform of a sequence by the core of the sequence characterizes regular complex matrices $A = (a_{nk})$ with $\lim_{n \to \infty} \sum_{k=0}^{\infty} |a_{nk}| = 1$ ([3, p. 149, Theorem 6.4, II]). In the non-Archimedean case, this condition characterizes regular matrices with $\overline{\lim_{n \to \infty}} \sup_{k \geq 0} |a_{nk}| = 1$. In the complex case, the core reflects the extent of divergence of a sequence. In the non-Archimedean case, presumably because of the absence of an effective notion of convexity, the core does not serve this purpose! In this case, the only proper K-convex subsets of K are spheres!

Chapter 5 is devoted to a study of the spaces $c_0(p)$ defined with respect to a positive, bounded sequence $p = \{p_k\}$ and in relation to any valued field K including $\mathbb{R}$ or $\mathbb{C}$. The family of spaces $c_0(p)$ includes Ganapathy Iyer's space of entire functions when $K = \mathbb{C}$ and its non-Archimedean analogue studied later by Raghunathan [13]. It is shown that in each member of the family, Schur's property holds, i.e., weak and strong convergence coincide when K is a non-Archimedean valued field, while this is the case when $K = \mathbb{R}$ or $\mathbb{C}$ if and only if $\lim_{k \to \infty} p_k = 0$. On the other hand, $c_0(p)$ is normable, in any case, if and only if $\inf p_k > 0$ and in this case, $c_0(p)$ could be identified with c_0. Section 5.3 concerns nuclearity of the space $c_0(p)$ when $K = \mathbb{R}$ or $\mathbb{C}$. Sufficient conditions for nuclearity of $c_0(p)$ are obtained in this case. When K is a non-Archimedean valued field, the workable analogue of a nuclear space seems to be a Schwartz space as defined by De Grande-De Kimpe. When K is such a field and is further spherically complete, we prove that $c_0(p)$ is a Schwartz space if and only if $\lim_{k \to \infty} p_k = 0$. In Section 5.5, we study $c_0(p)$ as a metric linear algebra following Srinivasan's study for $p_k = \frac{1}{k}$ (see [19, 20]).

In Chapter 6, we study about the spaces $\ell(p), c_0(p), c(p), \ell_\infty(p)$, when K is a complete, non-trivially valued, non-Archimedean field. We record briefly some facts relating to continuous duals of these spaces and matrix transformations between these spaces when K is such a field. We also record some more properties of the above sequence spaces. Finally, we study the algebra $(\ell_\alpha, \ell_\alpha)$, $\alpha \geq 1$, in the context of a convolution product and prove a Mercerian theorem supplementing Theorem 29.

In Chapter 7, we prove a characterization of the matrix class (ℓ_∞, c_0) similar to the one of Maddox [8], when K is a complete, non-trivially valued, non-Archimedean field. Summability matrices of type M in such a field K were first introduced by Rangachari and Srinivasan [14]. However, this definition does not seem to be suitable in the non-Archimedean setup. We introduce a more suitable definition of summability matrices of type M in K and make a detailed study of such matrices.

In the final Chapter 8, $K = \mathbb{R}$ or $\mathbb{C}$ or a complete, non-trivially valued, non-Archimedean field. We prove several Steinhaus-type theorems in K, pointing out differences between the classical and non-Archimedean cases.

The special feature of the present book is that we present a comparative study of the classical and non-Archimedean results in the study of several sequence spaces and summability theory. To the author's knowledge, this book is the first such attempt.

The author profusely thanks Prof. Dr. V.K. Srinivasan for writing a Foreword for this book. He also thanks Mr. Boopal Ethirajan for typing the manuscript.

Bibliography

[1] R.V. Andree and G.M. Petersen, Matrix methods of summation regular for p-adic valuations, *Proc. Amer. Math. Soc.*, 7 (1956), 250–253.

[2] R.C. Buck, A note on subsequences, *Bull. Amer. Math. Soc.*, 49 (1943), 898–899.

[3] R.G. Cooke, *Infinite matrices and sequence spaces*, Macmillan, 1950.

[4] J.A. Fridy, Properties of absolute summability matrices, *Proc. Amer. Math. Soc.*, 24 (1970), 583–585.

[5] J.A. Fridy, Summability of rearrangements of sequences, *Math. Z.*, 143 (1975), 187–192.

[6] V. Ganapathy Iyer, On the space of integral functions I, *J. Indian Math. Soc.*, 12 (1948), 13–30.

[7] T.A. Keagy, Matrix transformations and absolute summability, *Pacific J. Math.*, 63 (1976), 411–415.

[8] I.J. Maddox, A Tauberian theorem for subsequences, *Bull. London Math. Soc.*, 2 (1970), 63–65.

[9] A.F. Monna, Sur le théorème de Banach-Steinhaus, *Indag. Math.*, 25 (1963), 121–131.

[10] P.N. Natarajan, The Steinhaus theorem for Toeplitz matrices in non-Archimedean fields, Comment. *Math. Prace Mat.*, 20 (1978), 417–422.

[11] P.N. Natarajan, Sequence spaces and matrix transformations over valued fields, Ph.D. thesis, University of Madras, June, 1980.

[12] P.N. Natarajan, Criterion for regular matrices in non-Archimedean fields, *J. Ramanujan Math. Soc.*, 6 (1991), 185–195.

[13] T.T. Raghunathan, On the space of entire functions over certain non-Archimedean fields, *Boll. Un. Mat. Ital.*, 1 (1968), 517–526.

[14] M.S. Rangachari and V.K. Srinivasan, Matrix transformations in non-Archimedean fields, *Indag. Math.*, 26 (1964), 422–429.

[15] J.B. Roberts, Matrix summability in *F*-fields, *Proc. Amer. Math. Soc.*, 8 (1957), 541–543.

[16] I. Schur, Über lineare Transforamtionen in der Theorie der unendlichen Reihen, *J. Reine Angew Math.*, 151 (1921), 79–111.

[17] D. Somasundaram, Some properties of *T*-matrices over non-Archimedean fields, *Publ. Math. Debrecen*, 21 (1974), 171–177.

[18] V.K. Srinivasan, On certain summation processes in the *p*-adic field, *Indag. Math.*, 27 (1965), 319–325.

[19] V.K. Srinivasan, On the ideal structure of the algebra of integral functions, *Proc. Nat. Inst. Sci. India Part A* 31 (1965), 368–374.

[20] V.K. Srinivasan, On the ideal structure of the algebra of all entire functions over complete non-Archimedean valued fields, *Arch. Math.* 24 (1973), 505–512.

Chapter 1

Preliminaries

This chapter is intended to present the background for the succeeding chapters. It aims at making the present book self-contained to the extent possible and contains all basic definitions and results of a general nature. For details leading to well-known results, special references are given at appropriate places. Knowledge of basic concepts and results in general topology and algebra is assumed.

1.1 Valuation and the topology induced by it

Definition 1. *A mapping $|\cdot| : K \to \mathbb{R}$, from a field K to the field $\mathbb{R}$ of real numbers, is called a valuation of rank 1 if*

(i) *$|\alpha| \geq 0$; $|\alpha| = 0$ if and only if $\alpha = 0$;*

(ii) *$|\alpha\beta| = |\alpha||\beta|$;*

    and

(iii) *$|\alpha + \beta| \leq |\alpha| + |\beta|$ (triangle inequality),*

    where $\alpha, \beta \in K$.

    The valuation is said to be non-Archimedean if the following stronger form of (iii) holds:

(i') *$|\alpha + \beta| \leq \max(|\alpha|, |\beta|)$ (stronger triangle inequality),*

$\alpha, \beta \in K$. *Otherwise, the valuation is said to be Archimedean.*

In this book, we deal with only valuations of rank 1 as given in Definition 1 and so omit the phrase "of rank 1" in the sequel. However, we note that a valuation of any rank on the field K can be defined as having its range in an ordered group. A field K, with a valuation $|\cdot|$ defined on it, is called a valued field. A valuation $|\cdot|$ on K induces, in a natural way, a metric d on K defined by

$$d(x, y) = |x - y|, \ x, y \in K. \tag{1.1}$$

If $|\cdot|$ is a non-Archimedean valuation on K, d satisfies the following stronger triangle inequality

$$d(x, y) \le \max(d(x, z), d(z, y)), \ x, y, z \in K. \tag{1.2}$$

On this account, d is called an ultrametric on K. A metric space (X, d), where d is an ultrametric, is called an ultrametric space.

Topological concepts on the valued field K such as open spheres, closed spheres, Cauchy sequences, convergent sequences, etc., can be defined as usual using the metric (1.1). We note that when the metric d on a metric space (X, d) satisfies (1.2), any sphere is both open and closed and hence the topology on X is zero-dimensional. In particular, the topology on a non-Archimedean valued field has this property.

Definition 2. *A valued field K with valuation $|\cdot|$ is said to be complete with respect to $|\cdot|$ if it is complete with respect to the metric d defined by (1.1).*

A valuation on a field K can be completed in the sense of the following.

Theorem 1. *[1, p. 33, Theorem 1.3] Given a valued field K with valuation $|\cdot|$, there exists a field $\tilde{K} \supset K$, unique up to isometric isomorphism, such that $\tilde{K}$ is complete with respect to a valuation extending $|\cdot|$ and K is dense in $\tilde{K}$.*

$\tilde{K}$ *is called the completion of K. In this context, it is to be noted that if $|\cdot|$ is a non-Archimedean valuation on K, then so is the extended valuation on $\tilde{K}$.*

Denoting the image of K in $\mathbb{R}$ under the valuation $|\cdot|$ by $|K|$, we have the following.

Theorem 2. *[1, p. 33, Theorem 1.4] If $|\cdot|$ is a non-Archimedean valuation on K and $\tilde{K}$ is the completion of K, $|K| = |\tilde{K}|$, the extended valuation also being denoted by $|\cdot|$.*

Often a non-Archimedean valuation is called an ultrametric valuation and a non-Archimedean valued field K is called an ultrametric field.

In the case of non-Archimedean valuations, a stronger version of the notion of completeness turns out to be more suitable. A collection of subsets $\{A_\mu\}$, $\mu \in I$, is called a "nest," if for $\mu_1, \mu_2 \in I$, either $A_{\mu_1} \subset A_{\mu_2}$ or $A_{\mu_2} \subset A_{\mu_1}$. Without loss of generality, when I is countable, we say that $\{A_i\}$, $i \in I$, is a "nested sequence" if $A_1 \supset A_2 \supset A_3 \supset \dots$.

Definition 3. *A field K with a non-Archimedean valuation is said to be "spherically complete" if every nest of closed spheres (in the metric of the valuation) has a non-empty intersection.*

In view of Cantor's theorem for metric spaces, it follows that if K is spherically complete, then it is also complete. However, there exist non-Archimedean valued fields which are complete but not spherically complete ([12, p. 81, Example 4]).

Many equivalent formulations of the notion of spherical completeness are known [12] but they are not listed here as they are not relevant to the present book.

1.2 Kinds of valuations

A valuation $|\cdot|$ on a field K is said to be trivial if $|\alpha| = 0$ for $\alpha = 0$ and $|\alpha| = 1$ for $\alpha \neq 0$. It is said to be non-trivial otherwise, i.e., if there exists

$\alpha \in K$, $\alpha \neq 0$ such that $|\alpha| < 1$. If the valuation on K is non-trivial, K is said to be non-trivially valued. Otherwise, K is said to be trivially valued.

Unless otherwise stated, K stands for a complete, non-trivially valued, non-Archimedean field.

We note that $|K - \{0\}|$ is a multiplicative subgroup of the positive reals $(\mathbb{R}^+, .)$. Thus $|K - \{0\}|$ is either a discrete (and hence an infinite cyclic) subgroup or a subgroup which is dense in the topology of $(\mathbb{R}^+, .)$. The classification of a valuation depends on this fact.

Definition 4. *The valuation $|\cdot|$ on K is said to be "discrete" (equivalently, K is discretely valued) if $|K - \{0\}|$ is a discrete subgroup of $(\mathbb{R}^+, .)$. It is called "dense" (equivalently, K is densely valued) if $|K - \{0\}|$ is a dense subgroup of $(\mathbb{R}^+, .)$.*

For convenience of later references, we state here the following theorem.

Theorem 3. *[12, p. 35, Theorem 3] A non-Archimedean discretely valued, complete field is spherically complete.*

Example 1. *The usual absolute value defined on $\mathbb{R}$ and $\mathbb{C}$ is a valuation. $\mathbb{R}$ and $\mathbb{C}$ are complete under this valuation, which is Archimedean. We also note that this valuation is dense.*

The most important example of a valuation which is relevant to the present book is the following.

Example 2. *Let p be any fixed prime number and ρ be any real number such that $0 < \rho < 1$. For any $x \in Q$ (the field of rational numbers), write*

$$x = p^n \frac{a}{b},$$

where $a, b, n \in \mathbb{Z}$ (the ring of integers), p does not divide a, b.
Define

$$|x|_p = \rho^n \quad and \quad |0|_p = 0.$$

It is now easy to check that $|\cdot|_p$ is a non-Archimedean valuation on Q. It is called the p-adic valuation on Q. Q is not, however, complete with respect to this valuation. The completion of Q in the sense of Theorem 1 with respect to $|\cdot|_p$ is denoted by Q_p and it is called the p-adic field. It now follows from Theorem 2 that $|Q_p|_p = |Q|_p$ and so Q_p is discretely valued. In view of Theorem 3, Q_p, being complete and discretely valued, is spherically complete.

In this context, it is worthwhile to note the only valuations on Q are

(i) the trivial valuation;

(ii) a power $|\cdot|^\alpha$ of the usual absolute value, where $0 < \alpha \leq 1$;

    and

(iii) the p-adic valuation.

The non-Archimedean valuations on Q are, therefore, those given by (i) and (iii).

For an example of a densely valued, non-Archimedean field, see ([5, p. 239, Exercise 2]).

Definition 5. *Let K be a field with a non-Archimedean valuation $|\cdot|$. The subring*

$$\mathbb{V} = \{\alpha \in K : |\alpha| \leq 1\}$$

is called the valuation ring associated with the valuation $|\cdot|$.

$$\mathbb{P} = \{\alpha \in K : |\alpha| < 1\}$$

is an ideal which is maximal in $\mathbb{V}$. Moreover, $\mathbb{P}$ is the unique maximal ideal of $\mathbb{V}$. The field $\mathbb{V}/\mathbb{P}$ is called the residue class field of K associated with the valuation $|\cdot|$.

In this context, we mention the following useful result.

Theorem 4. *[12, p. 23, Theorem 1 and p. 25. Corollary] The following statements are equivalent:*

 (*i*) *K is locally compact;*

 (*ii*) $\mathbb{V}$ *is compact;*

   and

 (*iii*) *the valuation on K is discrete, K is complete with respect to this valuation, and the residue class field is finite.*

Noting that the residue class field of Q_p is $\mathbb{Z}_p$, the field of integers modulo p, it follows by Theorem 4, that Q_p is locally compact.

1.3   Normed linear spaces

Definition 6. *Let X be a linear space over a non-trivially valued, non-Archimedean field K with valuation $|\cdot|$. A mapping $\|\cdot\| : X \to \mathbb{R}$ is called a non-Archimedean norm if*

 (*i*) $\|x\| \geq 0$*;* $\|x\| = 0$ *if and only if* $x = 0$*;*

 (*ii*) $\|\alpha x\| = |\alpha|\|x\|$*;*

   and

 (*iii*) $\|x + y\| \leq \max(\|x\|, \|y\|)$*,*

   where $x, y \in X$ and $\alpha \in K$.

X, along with a non-Archimedean norm $\|\cdot\|$ defined on it, is called a non-Archimedean normed linear space. Defining

$$\|X\| = \{\|x\| : x \in X\},$$

and

$$|K| = \{|\alpha| : \alpha \in K\},$$

the sets $\|X\|$, $|K|$ may not, in general, be related by any inclusion. There may not exist unit vectors in X, i.e., $x \in X$ such that $\|x\| = 1$. However, the inclusion $\|X\| \subset |K|$ ensures the existence of unit vectors in X and implies the reverse inclusion so that $\|X\| = |K|$. Whenever there is necessity, let us assume that $\|X\| = |K|$.

The following result also holds with $|\cdot|$ in the place of $\|\cdot\|$ when $X = K$, a non-Archimedean valued field.

Theorem 5. *[12, p. 5, Theorem 2] If X is a non-Archimedean normed linear space and if $\|x\| > \|y\|$, then $\|x + y\| = \|x\|$.*

A non-Archimedean normed linear space is an ultrametric space, as explained in Section 1.1, with the metric defined by

$$d(x, y) = \|x - y\|, \ x, y \in X.$$

The topology on X is the one induced by this metric and the topological properties of a non-Archimedean normed linear space mentioned in the sequel are in the sense of this topology. Trivially, a non-Archimedean valued field is a non-Archimedean normed linear space over itself.

The following results, which will be used in the sequel, quite often without explicit mention, are easily proved.

(i) Every point in an open (closed) sphere is a center of the sphere, i.e., $y \in S_\epsilon(x)$ $(C_\epsilon(x))$ implies $S_\epsilon(x) = S_\epsilon(y)$ $(C_\epsilon(x) = C_\epsilon(y))$, where

$$S_\epsilon(x) = \{z : \|z - x\| < \epsilon\},$$
$$C_\epsilon(x) = \{z : \|z - x\| \leq \epsilon\}.$$

(ii) If X is complete with respect to $\|\cdot\|$, a series $\sum_{n=0}^{\infty} x_n$ converges if and only if

$$\lim_{n \to \infty} x_n = 0.$$

(iii) A sequence $\{x_n\}$ is Cauchy if and only if

$$\lim_{n \to \infty} (x_{n+1} - x_n) = 0.$$

(iv) If $\lim\limits_{n\to\infty} x_n = x$, $x \neq 0$, then for sufficiently large n,

$$\|x_n\| = \|x\|.$$

If X, Y are non-Archimedean normed linear spaces over K, a linear mapping $T : X \to Y$ is said to be bounded if there exists $M > 0$ such that

$$\|T(x)\| \leq M\|x\|, \ x \in X.$$

The set $L(X, Y)$ of all bounded linear transformations from X to Y is a linear space. Defining, for $T \in L(X, Y)$,

$$\|T\| = \sup_{\substack{x \in X \\ x \neq 0}} \left\{ \frac{\|T(x)\|}{\|x\|} \right\}, \tag{1.3}$$

$\|\cdot\|$ is a non-Archimedean norm on $L(X, Y)$. It is easy to check that $L(X, Y)$ is complete with respect to the norm defined by (1.3) if Y is complete. If $Y = K$, $L(X, K)$ is called the continuous dual of X and is denoted by X^*. Elements of X^* are called bounded linear functionals on X.

Equivalent formulations of (1.3), familiar in the case of normed linear spaces over $\mathbb{R}$ or $\mathbb{C}$, are no more true when $\mathbb{R}$ or $\mathbb{C}$ is replaced by a field K with a non-Archimedean valuation (see [12, pp. 75–76]). However, if K is non-trivially valued, then the equivalence of

(i) continuity of a linear transformation T;

(ii) T is bounded;

and

(iii) its boundedness on the unit sphere, viz., there exists $M > 0$ such that $\|T(x)\| \leq M$ if $\|x\| \leq 1$,

is retained (see [12, p. 77, Lemma 1]).

Definition 7. *Let X be a linear space over a valued field K. By a paranorm on X, we mean a mapping $g : X \to \mathbb{R}$ such that* *(i)* $g(0) = 0;$

(ii) $g(x) = g(-x);$

(iii) $g(x + y) \leq g(x) + g(y);$

   and

(iv) $\lambda \to \lambda_0$, $g(x - x_0) \to 0$ *imply* $g(\lambda x - \lambda_0 x_0) \to 0$,

where $\lambda, \lambda_0 \in K$ *and* $x, x_0 \in X$. X, *with a paranorm* g, *is called a paranormed space. If, as usual, (iii) is replaced by the stronger triangle inequality*

(i') $g(x + y) \leq \max(g(x), g(y))$,

g is called a non-Archimedean paranorm on X *and* X *is called a non-Archimedean paranormed space. In this context, we note that*

$$d(x, y) = g(x - y)$$

defines a pseudo-metric on X.

A linear functional A on a paranormed space is continuous if and only if

$$\|A\|_M = \sup_{x \in X} \left\{ |A(x)| : g(x) \leq \frac{1}{M} \right\} < \infty$$

for some $M > 1$.

1.4   *K*-convexity and locally *K*-convex spaces

A topological linear space X over a valued field K is a linear space with a Hausdorff topology which is compatible with the linear space structure, i.e., the mappings,

$$(x, y) \to x + y : X \times X \to X$$

and

$$(\alpha, x) \to \alpha x : K \times X \to X$$

are continuous. A subset S of X is said to be balanced if $\alpha S \subset S$, whenever $|\alpha| \leq 1$, $\alpha \in K$. A topological linear space is called non-Archimedean if there exists a filter base $\mathscr{U}$ of balanced sets for the filter of all neighborhood of 0 such that for all $U \in \mathscr{U}$, $U + U \subset U$.

Definition 8. *[12, p. 55] Let X be a topological linear space over a non-trivially valued, non-Archimedean field K. A subset S of X is said to be absolutely K-convex if S is a $\mathbb{V}$-module, i.e., $\mathbb{V}S + \mathbb{V}S \subset S$. S is said to be K-convex if S is absolutely K-convex or a translate of an absolutely K-convex set.*

In this connection, we note that any sphere in a non-Archimedean normed linear space is K-convex.

Treating K as a topological linear space over itself, the following result, which is easily proved, is noteworthy.

Theorem 6. *[12, p. 27, Exercise 1.11] The only proper K-convex subsets of K are spheres.*

The following result gives some equivalent formulations for K-convexity.

Theorem 7. *[12, p. 56] X is a topological linear space over K.*

(i) *$S \subset X$ is K-convex if and only if for all $\lambda, \mu, \gamma \in K$ with $\lambda + \mu + \gamma = 1$, we have $\lambda x + \mu y + \gamma z \in S$ for all $x, y, z \in S$.*

(ii) *If the characteristic of $\mathbb{V}/\mathbb{P}$, the residue class field of K, is not 2, then $S \subset X$ is K-convex if and only if for all $\lambda \in \mathbb{V}$ and for all $x, y \in S$, $\lambda x + (1 - \lambda)y \in S$, i.e., non-Archimedean K-convexity is analogous to classical convexity when characteristic of $\mathbb{V}/\mathbb{P} \neq 2$.*

Definition 9. *A linear space X over K is said to be locally K-convex if there exists a fundamental system of K-convex neighborhoods of 0.*

We note that X is a non-Archimedean topological linear space if and only if K is non-Archimedean valued and X is locally K-convex (see [9]; [12, p. 61, Exercise 2.15]).

As in the classical case, the topology of a linear ultrametric space, which is incidentally a locally K-convex topological linear space, is given by an increasing sequence of non-Archimedean seminorms (for details, see van Tiel [16]).

Definition 10. *A non-Archimedean topological linear space X over K is said to be normable if the topology of X is defined by a non-Archimedean norm.*

The following theorem gives the criterion for the normability of a non-Archimedean topological linear space over K.

Theorem 8. *[9, p. 360] A necessary and sufficient condition for the topology of a non-Archimedean topological linear space X over K to be equivalent to one defined by a non-Archimedean norm, i.e., for X to be normable, is that there exists a bounded K-convex neighborhood of 0.*

In the context of a sequence space, the concept of a "step space," defined as follows, is useful.

Definition 11. *[2, p. 53] Let E be a sequence space with entries of the sequences in a valued field K. Let I be a subsequence of positive integers. Then*

$$\lambda_I = \{\{x_k\} \in E : x_k = 0, k \notin I\}$$

is called a step space of E corresponding to I.

1.5 Topological algebras

Definition 12. *Let X be an algebra over K. X is called a normed algebra if there exists a norm $\|\cdot\|$ on X such that $\|xy\| \leq \|x\|\|y\|$ for all $x, y \in X$. If X is complete with respect to the norm, X is called a Banach algebra.*

The following definitions are needed in the sequel.

Definition 13. *[11, p. 167] X is a topological linear space over a valued field K. X is called a topological algebra if*

(i) *X is an algebra;*

(ii) *X is locally convex (locally K-convex);*

   and

(iii) *the mapping $(x, y) \to xy$ is continuous in each of the variables provided the other variable is fixed.*

Definition 14. *[11, p. 155] If X is a commutative topological algebra and $x \in X$, x is said to be quasi-invertible with y as quasi-inverse if*

$$x + y + xy = 0.$$

Definition 15. *[11, p. 157] An ideal I of a commutative algebra X is said to be regular if there exists $u \in X$ such that*

$$ux - x \in I \quad \text{for every } x \in X.$$

1.6   Summability methods

ℓ_∞ denotes the set of all bounded sequences with entries in K, where K is a complete, non-trivially valued field. Defining, for $x = \{x_k\} \in \ell_\infty$, the norm of x by

$$\|x\| = \sup_{k \geq 0} |x_k|, \tag{1.4}$$

ℓ_∞ is seen to be a Banach space when $K = \mathbb{R}$ or $\mathbb{C}$ and a non-Archimedean Banach space when K is non-Archimedean valued. c, c_0 denote, respectively, the sets of convergent sequences and null sequences. With respect to the norm defined by (1.4), c, c_0 are closed subspaces of ℓ_∞.

Definition 16. *If $A = (a_{nk})$, $a_{nk} \in K$, $n, k = 0, 1, 2, \ldots$ is an infinite matrix, by the A-transform Ax (or $A(x)$) of a sequence $x = \{x_k\}$, $x_k \in K$, $k = 0, 1, 2, \ldots$, we mean the sequence $\{(Ax)_n\}$, where*

$$(Ax)_n = \sum_{k=0}^{\infty} a_{nk} x_k, \ n = 0, 1, 2, \ldots,$$

assuming that the series on the right converge. The sequence $x = \{x_k\}$ is said to be summable by the matrix summability method A or in short, A-summable, to s if

$$(Ax)_n \to s, \ n \to \infty.$$

Definition 17. *If X, Y are sequence spaces with elements whose entries are in K and if $A = (a_{nk})$, $a_{nk} \in K$, $n, k = 0, 1, 2, \ldots$ is an infinite matrix, A is said to transform X to Y if, whenever $x = \{x_k\} \in X$, $(Ax)_n$ is defined, $n = 0, 1, 2, \ldots$ and the sequence $\{(Ax)_n\} \in Y$. In this case, we write*

$$A \in (X, Y).$$

Definition 18. *Let $A = (a_{nk})$, $a_{nk} \in K$, $n, k = 0, 1, 2, \ldots$ be an infinite matrix. If $A \in (c, c)$, A is said to be convergence-preserving or conservative. If, further,*

$$\lim_{n \to \infty} (Ax)_n = \lim_{k \to \infty} x_k,$$

A is said to be a regular matrix or a Toeplitz matrix. In this case, we write $A \in (c, c; P)$, the letter P standing for "preservation of limits."

When $K = \mathbb{R}$ of $\mathbb{C}$, necessary and sufficient conditions for $A = (a_{nk})$ to be conservative or regular, in terms of the entries of the matrix, are very well known (see [4]).

When K is a complete, non-trivially valued, non-Archimedean field, we have the following result.

Theorem 9 (see [10], [13]). *When K is a complete, non-trivially valued, non-Archimedean field, $A = (a_{nk})$ is conservative if and only if*

(i) $\sup\limits_{n,k} |a_{nk}| < \infty$;

(ii) $\lim\limits_{n\to\infty} a_{nk} = \delta_k$ *exists,* $k = 0, 1, 2, \ldots$;

and

(iii) $\lim\limits_{n\to\infty} \sum\limits_{k=0}^{\infty} a_{nk} = \delta$ *exists.*

Further, A is regular if and only if (i)-(iii) hold with $\delta_k = 0$, $k = 0, 1, 2, \ldots$ and $\delta = 1$.

Definition 19. $A = (a_{nk})$ *is called a Schur matrix if* $A \in (\ell_\infty, c)$.

Definition 20 (see e.g., [8], [15]). *If* $x_k \in K$, $k = 0, 1, 2, \ldots$, *the following sequence spaces are defined. If* $p = \{p_k\}$ *is a sequence of positive real numbers,*

$$c_0(p) = \{x = \{x_k\} : |x_k|^{p_k} \to 0, k \to \infty\};$$

$$c(p) = \{x = \{x_k\} : |x_k - s|^{p_k} \to 0, k \to \infty \text{ for some } s \in K\};$$

$$\ell_\infty(p) = \left\{x = \{x_k\} : \sup_{k\geq 0} |x_k|^{p_k} < \infty\right\};$$

and

$$\ell(p) = \left\{x = \{x_k\} : \sum_{k=0}^{\infty} |x_k|^{p_k} < \infty\right\}.$$

It is to be noted that when $p_k = \alpha$, $k = 0, 1, 2, \ldots$, *these spaces reduce to* $c_0, c, \ell_\infty, \ell_\alpha$, *respectively, where*

$$\ell_\alpha = \left\{x = \{x_k\} : \sum_{k=0}^{\infty} |x_k|^{\alpha} < \infty\right\}.$$

It is easily seen that the above spaces are linear spaces if and only if

$$\sup_{k\geq 0} p_k < \infty. \tag{1.5}$$

In the sequel, we shall suppose that the sequence $\{p_k\}$ *satisfies (1.5). When* $p_0 = 1$, $p_k = \frac{1}{k}$, $k = 1, 2, \ldots$, $c_0(p)$ *can be identified with the space of entire functions on* K *taking values in* K. *It may be recalled ([3], [14]) that in this* *special case, $c_0(p)$ is a non-normable metric linear space in which weak and strong convergence coincide. It is clear that*

$$c_0(p) \subset c_0 \subset \ell_\infty.$$

We note that

$$c_0(p) = c_0, c(p) = c, \ell_\infty(p) = \ell_\infty$$

if and only if

$$\inf_{k \geq 0} p_k > 0. \tag{1.6}$$

The paranorm on $\ell_\infty(p)$ is given by

$$g(x) = \sup_{k \geq 0} |x_k|^{p_k/H}, H = \max(1, \sup_{k \geq 0} p_k). \tag{1.7}$$

In fact,

$$d(x, y) = g(x - y)$$

defines a metric on $\ell_\infty(p)$ with respect to which it is a complete metric space. However, $\ell_\infty(p)$ is a metric linear space if and only if (1.6) is satisfied. $c_0(p)$, $c(p)$ are closed metric linear subspaces of $\ell_\infty(p)$ with respect to the paranorm defined by (1.7). The paranorm on $\ell(p)$ is defined by

$$g(x) = \left(\sum_{k=0}^{\infty} |x_k|^{p_k}\right)^{\frac{1}{H}}, H = \max(1, \sup_{k \geq 0} p_k). \tag{1.8}$$

In this case too,

$$d(x, y) = g(x - y)$$

defines a metric on $\ell(p)$ with respect to which $\ell(p)$ is a complete metric linear space.

The following notion is useful in the study of matrix transformations.

Definition 21. *If E is a set of sequences with entries in a valued field K, its (generalized) Köthe-Toeplitz dual, denoted by $E^\times$, is defined as*

$$E^\times = \left\{ x = \{x_k\} : \sum_{k=0}^{\infty} x_k y_k \text{ converges for all } y = \{y_k\} \in E \right\}.$$

The following details are easily verified:

$K = \mathbb{R}$ or $\mathbb{C}$	K is a complete, non-trivially valued, non-Archimedean field
$c_0^\times(p) = M_0(p)$ [8]	$c_0^\times(p) = \ell_\infty(p)$
$\ell_\infty^\times(p) = M_\infty(p)$ [7]	$\ell_\infty^\times(p) = c_0(p)$

where

$$M_0(p) = \bigcup_{N>1} \left\{ \{x_k\} : \sum_{k=0}^{\infty} |x_k| N^{-1/p_k} < \infty \right\},$$

$$M_\infty(p) = \bigcap_{N>1} \left\{ \{x_k\} : \sum_{k=0}^{\infty} |x_k| N^{1/p_k} < \infty \right\},$$

$c_0^\times(p)$, $\ell_\infty^\times(p)$ denoting the Köthe-Toeplitz duals of $c_0(p)$, $\ell_\infty(p)$ respectively.

Definition 22. *If E is a set of sequences with entries in a valued field K, its Köthe-Toeplitz dual (called the α-dual by Köthe [6, p. 405]), denoted by E^+ is defined as*

$$E^+ = \left\{ x = \{x_k\} : \sum_{k=1}^{\infty} |x_k y_k| < \infty \text{ for all } y = \{y_k\} \in E \right\}.$$

It may be noted that $E^+ \subset E^\times$.

Bibliography

[1] G. Bachman. *Introduction to p-adic numbers and valuation theory*. Academic Press, 1964.

[2] N. De Grande-De Kimpe and W.B. Robinson. Compact maps and embeddings from an infinite type power series space to a finite power series space. *J. Reine Angew. Math.*, 293/294:52–61, 1977.

[3] V. Ganapathy Iyer. On the space of integral functions I. *J. Indian Math. Soc.*, 12:13–30, 1948.

[4] G.H. Hardy. *Divergent Series*. Oxford, 1949.

[5] N. Jacobson. *Lectures in abstract algebra*. Vol. III, van Nostrand, 1964.

[6] G. Köthe. *Topological vector spaces I*. Springer, 1969.

[7] C.G. Lascarides and I.J. Maddox. Matrix transformations between some classes of sequences. *Proc. Cambridge Philos. Soc.*, 68:99–104, 1970.

[8] I.J. Maddox. Continuous and Köthe-Toeplitz duals of certain sequence spaces. *Proc. Cambridge Philos. Soc.*, 65:431–435, 1969.

[9] A.F. Monna. Espaces vectoriels topologiques sur un corps valué. *Indag. Math.*, 24:351–367, 1962.

[10] A.F. Monna. Sur le théorème de Banach-Steinhaus. *Indag. Math.*, 25:121–131, 1963.

[11] M.A. Naimark. *Normed Algebras*. Noordhoff, 1972.

[12] L. Narici, E. Beckenstein and G. Bachman. *Functional Analysis and Valuation Theory*. Marcel Dekker, 1971.

[13] P.N. Natarajan. Criterion for regular matrices in non-archimedean fields. *J. Ramanujan Math. Soc.*, 6:185–195, 1991.

[14] T.T. Raghunathan. On the space of entire functions over certain non-archimedean fields. *Boll. Un. Mat. Ital.*, 1:517–526, 1968.

[15] S. Simons. The sequence space $\ell(p_\gamma)$ and $m(p_\gamma)$. *Proc. London Math. Soc.*, 15:422–436, 1965.

[16] J. van Tiel. Espaces localement K-convexes I-III. *Indag. Math.*, 27:249–258; 259–272; 273–289, 1965.

Chapter 2

On Certain Spaces Containing the Space of Cauchy Sequences

2.1   Introduction

In this chapter, the field K may be $\mathbb{R}$ or $\mathbb{C}$ or a non-trivially valued, non-Archimedean field. In the relevant context, we explicitly mention which field is chosen. In the absence of such an explicit mention, the field K can be one of these. The sequence spaces mentioned in the title are those introduced by:

Definition 23. *(see [3], [4], [5]) The sequence $x = \{x_k\}$, $x_k \in K$, $k = 0, 1, 2, \ldots$ is said to belong to the space Λ_r, $r = 1, 2, \ldots$, if $\{x_k\} \in \ell_\infty$ and*

$$|x_{k+r} - x_k| \to 0, \ k \to \infty.$$

It is clear that any Cauchy sequence is in $\bigcap_{r=1}^{\infty} \Lambda_r$ so that each Λ_r is a sequence space containing the space $\mathscr{C}$ of Cauchy sequences. However, it may be noted that

$$\mathscr{C} \subsetneqq \bigcap_{r=1}^{\infty} \Lambda_r, \text{ when } K = \mathbb{R} \text{ or } \mathbb{C},$$

while,

$$\mathscr{C} = \bigcap_{r=1}^{\infty} \Lambda_r, \text{ when } K \text{ is a non-trivially valued,}$$

non-Archimedean field. The first assertion is justified by the sequence $\{x_k\}$, $x_k = 1 + \frac{1}{2} + \cdots + \frac{1}{k+1}$, $k = 0, 1, 2, \ldots$, which belongs to each Λ_r but not to

$\mathscr{C}$. On the other hand, if K is a non-trivially valued, non-Archimedean field, $\Lambda_1 = \mathscr{C}$ so that $\bigcap_{r=1}^{\infty} \Lambda_r \subset \mathscr{C}$; but $\mathscr{C} \subset \bigcap_{r=1}^{\infty} \Lambda_r$ so that $\mathscr{C} = \bigcap_{r=1}^{\infty} \Lambda_r$, which is the second assertion. Though Λ_r do not form a tower between $\mathscr{C}$ and ℓ_{∞}, they can be deemed to reflect the measure of non-Cauchy nature of sequences contained in them. By the way, it is easy to prove that $\Lambda_r \subset \Lambda_s$ if and only if s is a multiple of r and that $\Lambda_r \cap \Lambda_{r+1} = \Lambda_1$.

As the main object of the present chapter is to study Steinhaus-type theorems in relation to matrix transformations, it is useful to study the nature of the location of sequences of 0's and 1's in these spaces Λ_r. In the first instance, we note that a sequence of 0's and 1's is in Λ_r if and only if it is periodic, with period r, after a certain stage. Consequently, any sequence of 0's and 1's is in $\ell_{\infty} - \bigcup_{r=1}^{\infty} \Lambda_r$ if and only if it is non-periodic. In this context, we note that the sequences

$$e_i^{(r)} = \{e_{ik}^{(r)}\}_{k=0}^{\infty} \tag{2.1}$$

$$= \left\{ \underbrace{1, 1, \ldots, 1}_{i}, \underbrace{0, 0, \ldots, 0}_{r-i}, \underbrace{1, 1, \ldots, 1}_{i}, \underbrace{0, 0, \ldots, 0}_{r-i}, \ldots \right\},$$

$$i = 1, 2, \ldots, r,$$

have a role to play in the structure of Λ_r, though there are sequences of 0's and 1's in Λ_r which are not necessarily of the form (2.1),

$$\text{e.g.,} \quad \left\{ 1, 0, 1, \underbrace{0, 0, \ldots, 0}_{r-3}, 1, 0, 1, \underbrace{0, 0, \ldots, 0}_{r-3}, \ldots \right\}.$$

Their role is brought out by the following result.

Theorem 10. *An infinite matrix* $A = (a_{nk})$, $a_{nk} \in K$, $n, k = 0, 1, 2, \ldots$, *for which* $\lim_{n \to \infty} a_{nk}$ *exists*, $k = 0, 1, 2, \ldots$, *sums every sequence of 0's and 1's in* Λ_r *if and only if it sums the sequences in (2.1).*

Proof. It suffices to prove the sufficiency part of the theorem. From the observation made earlier, a sequence $x = \{x_k\} \in \Lambda_r$, $x_k = 0$ or 1, $k = 0, 1, 2, \ldots$ is necessarily periodic, with period r, after a certain stage. Let $\{x_k\}$ be periodic with period r for $k \geq k_0$, where without loss of generality, we can suppose that $k_0 \equiv 0 \pmod{r}$. The system of linear equations in the r unknowns λ_i, $i = 1, 2, \ldots, r$, viz.,

$$x_{k_0+k} = \sum_{i=1}^{r} \lambda_i e_{ik}^{(r)}, \ k = 0, 1, \ldots, (r-1)$$

has a unique solution, since the determinant of the matrix $(e_{ik}^{(r)})$ is unity. Noting that

$$x_{k+pr} = x_k,$$
$$e_{i,k+pr}^{(r)} = e_{ik}^{(r)}, \ k \geq k_0, p = 0, 1, 2, \ldots,$$

$$
\begin{aligned}
\sum_{k=0}^{\infty} a_{nk} x_k &= \sum_{k=0}^{k_0-1} a_{nk} x_k + \sum_{k=k_0}^{\infty} a_{nk} x_k \\
&= \sum_{k=0}^{k_0-1} a_{nk} x_k + \sum_{p=0}^{\infty} \sum_{k=k_0+pr}^{k_0+(p+1)r-1} a_{nk} \left(\sum_{i=1}^{r} \lambda_i e_{ik}^{(r)} \right) \\
&= \sum_{k=0}^{k_0-1} a_{nk} x_k + \sum_{i=1}^{r} \lambda_i \left(\sum_{p=0}^{\infty} \sum_{k=k_0+pr}^{k_0+(p+1)r-1} a_{nk} e_{ik}^{(r)} \right) \\
&= \sum_{k=0}^{k_0-1} a_{nk} x_k + \sum_{i=1}^{r} \lambda_i \left(\sum_{k=k_0}^{\infty} a_{nk} e_{ik}^{(r)} \right),
\end{aligned}
$$

from which it is clear that $\{x_k\}$ is summable A, since $e_i^{(r)}$, $i = 1, 2, \ldots, r$ are summable A and $\lim_{n \to \infty} a_{nk}$ exists, $k = 0, 1, 2, \ldots$. $\qquad\square$

2.2   Summability of sequences of 0's and 1's

In view of Schur's version of the Steinhaus theorem [9], viz., given a regular matrix A, there exists a sequence of 0's and 1's not summable A, the following questions naturally arise:

(1) Given a regular matrix A, does there exist a sequence of 0's and 1's in $\ell_\infty - \bigcup_{r=1}^{\infty} \Lambda_r$, i.e., non-periodic, which is not summable A?

(2) Given a regular matrix A, does there exist a sequence of 0's and 1's in $\bigcup_{r=1}^{\infty} \Lambda_r$, i.e., periodic after a stage, which is not summable A?

The results given below provide the affirmative and negative answers in that order to the above questions.

Theorem 11. *Given a regular matrix* $A = (a_{nk})$, $a_{nk} \in K$, $n, k = 0, 1, 2, \dots$, *there exists a sequence of 0's and 1's in* $\ell_\infty - \bigcup_{r=1}^{\infty} \Lambda_r$, *which is not summable* A.

Proof. Case 1. $K = \mathbb{R}$ or $\mathbb{C}$.

Choose positive integers $n(1)$, $k(1)$ such that

$$\left| \sum_{k=0}^{k(1)} a_{n(1),k} \right| > \frac{7}{8},$$

$$\sum_{k=k(1)+1}^{\infty} |a_{n(1),k}| < \frac{1}{16}.$$

Now, positive integers $n(2)$, $k(2)$ can be chosen such that $n(2) > n(1)$, $k(2) > k(1)$,

$$\sum_{k=0}^{k(1)} |a_{n(2),k}| < \frac{1}{16},$$

$$\sum_{k=k(2)+1}^{\infty} |a_{n(2),k}| < \frac{1}{16}.$$

Then choose positive integers $n(3)$, $k(3)$ such that $n(3) > n(2)$, $k(3) > k(2)+1$,

$$\sum_{k=0}^{k(2)+1} |a_{n(3),k}| < \frac{1}{16},$$

$$\left| \sum_{k=k(2)+2}^{k(3)} a_{n(3),k} \right| > \frac{7}{8},$$

$$\sum_{k=k(3)+1}^{\infty} |a_{n(3),k}| < \frac{1}{16}.$$

More generally, given the positive integers $n(j)$, $k(j)$, $j \leq m-1$, if $m = 2p$, choose positive integers $n(m)$, $k(m)$ such that $n(m) > n(m-1)$, $k(m) > k(m-1)+m-2$,

$$\sum_{k=0}^{k(m-1)+m-2} |a_{n(m),k}| < \frac{1}{16},$$

$$\sum_{k=k(m)+1}^{\infty} |a_{n(m),k}| < \frac{1}{16};$$

if $m = 2p+1$, choose $n(m)$, $k(m)$ so that $n(m) > n(m-1)$, $k(m) > k(m-1)+m-2$,

$$\sum_{k=0}^{k(m-1)+m-2} |a_{n(m),k}| < \frac{1}{16},$$

$$\left| \sum_{k=k(m-1)+m-1}^{k(m)} a_{n(m),k} \right| > \frac{7}{8},$$

and

$$\sum_{k=k(m)+1}^{\infty} |a_{n(m),k}| < \frac{1}{16}.$$

Define the sequence $\{x_k\}$ as follows:

$$
\left.
\begin{aligned}
x_k &= 0, \ k(2p-1) < k \leq k(2p) \\
&= 1, \ k(2p) < k \leq k(2p+1)
\end{aligned}
\right\} , p = 1, 2, \ldots.
\tag{2.2}
$$

It is clear that $\{x_k\} \in \ell_\infty - \bigcup_{r=1}^{\infty} \Lambda_r$. Also,

$$
\begin{aligned}
|(Ax)_{n(2p+1)}| &\geq \left| \sum_{k=k(2p)+2p}^{k(2p+1)} a_{n(2p+1),k} \right| \\
&\quad - \sum_{k=0}^{k(2p)+2p-1} |a_{n(2p+1),k}| \\
&\quad - \sum_{k=k(2p+1)+1}^{\infty} |a_{n(2p+1),k}| \\
&> \frac{7}{8} - \frac{1}{16} - \frac{1}{16} \\
&= \frac{3}{4};
\end{aligned}
$$

$$
\begin{aligned}
|(Ax)_{n(2p)}| &\leq \sum_{k=0}^{k(2p-1)+2p-2} |a_{n(2p),k}| \\
&\quad + \sum_{k=k(2p)+1}^{\infty} |a_{n(2p),k}| \\
&< \frac{1}{16} + \frac{1}{16} \\
&= \frac{1}{8},
\end{aligned}
$$

$p = 1, 2, \ldots$, so that $\{(Ax)_n\} \notin c$, though $\{x_k\} \in \ell_\infty - \bigcup_{r=1}^{\infty} \Lambda_r$.

Case 2. K is a complete, non-trivially valued, non-Archimedean field. As in case 1, we choose sequences $\{n(m)\}$, $\{k(m)\}$ of positive integers for which $n(m) > n(m-1)$, $k(m) > k(m-1) + m - 2$ such that if $m = 2p$,

$$
\sup_{0 \leq k \leq k(m-1)+m-2} |a_{n(m),k}| < \frac{1}{16},
$$

$$
\sup_{k > k(m)} |a_{n(m),k}| < \frac{1}{16};
$$

if $m = 2p + 1$,

$$\sup_{0 \leq k \leq k(m-1)+m-2} |a_{n(m),k}| < \frac{1}{16},$$

$$\left| \sum_{k=k(m-1)+m-1}^{k(m)} a_{n(m),k} \right| > \frac{7}{8},$$

$$\sup_{k>k(m)} |a_{n(m),k}| < \frac{1}{16}.$$

Consider the sequence $x = \{x_k\}$ defined by (2.2). It is easy to check that

$$|(Ax)_{n(2p)}| < \frac{1}{8},$$

$$|(Ax)_{n(2p+1)}| > \frac{3}{4},$$

proving that $\{(Ax)_n\} \notin c$, completing the proof of the theorem.   □

Theorem 12. *There exists a regular matrix which sums all sequences of 0's and 1's in* $\bigcup_{r=1}^{\infty} \Lambda_r$.

Proof. Let $K = \mathbb{R}$ or $\mathbb{C}$ or $\mathbb{Q}_p$, the p-adic field for a prime p. Consider the infinite matrix

$$A \equiv (a_{nk}) = \begin{bmatrix} 1 & 0 & 0 & 0 & 0 & 0 & \cdots \\ 0 & \frac{1}{2} & \frac{1}{2} & 0 & 0 & 0 & \cdots \\ 0 & 0 & \frac{1}{3} & \frac{1}{3} & \frac{1}{3} & 0 & \cdots \\ \cdots & \cdots & \cdots & \cdots & \cdots & \cdots & \cdots \\ \cdots & \cdots & \cdots & \cdots & \cdots & \cdots & \cdots \end{bmatrix}.$$

It is clear that A is a regular matrix. For $x = \{x_k\}$,

$$(Ax)_n = \frac{x_n + x_{n+1} + \cdots + x_{2n-1}}{n}.$$

n has one of the forms λr, $\lambda r + 1$, ..., $\lambda r + (r-1)$, r being a fixed positive integer. Consider the sequence $e_1^{(r)}$ in (2.1). If

$n = \lambda r,$ $\qquad\qquad (Ae_1^{(r)})_n = \frac{\lambda}{\lambda r};$

$n = \lambda r + 1,$ $\qquad\quad (Ae_1^{(r)})_n = \frac{\lambda}{\lambda r + 1}; \ldots$

$n = \lambda r + (r-1),$ $\quad (Ae_1^{(r)})_n = \frac{\lambda + 1}{\lambda r + r - 1},$

which shows that

$$(Ae_1^{(r)})_n \to \frac{1}{r}, \ n \to \infty.$$

Similar computation shows that the A-transforms of $e_i^{(r)}$ converges to $\frac{i}{r}$, $i = 2, 3, \ldots, r$. This process could be repeated for any positive integer r. Thus, using Theorem 10, A sums all sequences of 0's and 1's in $\bigcup\limits_{r=1}^{\infty} \Lambda_r$, completing the proof. $\qquad\qquad\qquad\qquad\qquad\qquad\qquad\qquad\qquad\qquad\qquad\qquad\quad\square$

Remark 1. *It thus turns out that the probability of success in our search for a sequence of 0's and 1's not summable by a given regular matrix is more when we concentrate on non-periodic sequences than on sequences which are periodic after a stage. A complete picture of the situation, however, requires a characterization of regular matrices which fail to sum at least one sequence of 0's and 1's in $\bigcup\limits_{r=1}^{\infty} \Lambda_r$.*

2.3   Some structural properties of ℓ_∞

We now record some of the structural properties of ℓ_∞, vis-a-vis, the set of sequences of 0's and 1's. First, it can be shown that if K is locally compact (when $K = \mathbb{R}$ or $\mathbb{C}$, see [1] for details), the closed linear span of the set of all sequences of 0's and 1's in the supremum norm is ℓ_∞. The following result is an improvement of this assertion. Let $\mathscr{P}$ denote the set of all sequences of 0's and 1's in $\bigcup\limits_{r=1}^{\infty} \Lambda_r$, and $\mathscr{N}P$ the set of all non-periodic sequences of 0's and 1's.

Theorem 13. *The closed linear span of $\mathcal{N}P$ is ℓ_∞.*

Proof. It suffices to prove that periodic sequences of 0's and 1's, which are periodic from the beginning, are in the closed span of $\mathcal{N}P$. In fact, they are in the linear span of $\mathcal{N}P$. To this end, we show that any such sequence is the difference of two sequences in $\mathcal{N}P$. It is clear that any sequence of 0's and 1's, which converges to 0 or 1, can be expressed as the difference of two non-periodic sequences. Hence we shall take a divergent sequence $x = \{x_k\}$, $x_k = 0$ or 1, $x_{k+r} = x_k$. Let $\{d_i\}$ be a proper subsequence of positive integers such that $\{d_{i+1} - d_i\}$ increases to ∞. We first note that there are at least two consecutive entries, one of which is 0 while the other is 1, in between $x_{\lambda r}$ and $x_{(\lambda+1)r-1}$, $\lambda = 0, 1, 2, \ldots$. We now construct two sequences $x^{(1)}, x^{(2)}$ using x as follows:

$$x_k^{(1)} = x_k, (d_i - 1)r \le k < d_i r \text{ and}$$
$$d_{i-1}r \le k < (d_i - 1)r,$$

except when x_k, x_{k+1} are 0, 1 or 1, 0 respectively, the rank of the pair having been chosen already in each of the blocks of r terms. In the pair of terms chosen in each of the blocks, we change the pair 1, 0 or 0, 1 to 1, 1.

$$x_k^{(2)} = 0, \text{ when } x_k^{(1)} = x_k \text{ while } x_k^{(2)} = 1 \text{ if } x_k = 0 \text{ and } x_k^{(2)} = 0 \text{ if } x_k = 1$$

corresponding to k in the pairs chosen already. By construction,

$$x_k = x_k^{(1)} - x_k^{(2)},$$

where both $\{x_k^{(1)}\}$ and $\{x_k^{(2)}\}$ are non-periodic, since, for sufficiently large i, $x_{(d_i-1)r+j} \ne x_{(d_i-1)r+j-p}$ for a suitable j, $0 \le j \le r - 1$, j being chosen such that the term of this rank in the block of r terms has undergone a change in the chosen pair of terms, completing the proof of the theorem. $\qquad \square$

The following is an important result due to Schur [9]. We prove the theorem for the sake of completeness.

Theorem 14. *When $K = \mathbb{R}$ or $\mathbb{C}$, any matrix which sums all sequences of 0's and 1's is necessarily a Schur matrix, i.e., A sums all bounded sequences.*

Proof. It suffices to prove the theorem for the case $K = \mathbb{R}$. We first note that when A sums all sequences of 0's and 1's, it sums all sequences of 0's, 1's and -1's. We, then, note that

$$\sum_{k=0}^{\infty} |a_{nk}| < \infty, \ n = 0, 1, 2, \ldots.$$

For, if $\sum\limits_{k=0}^{\infty} |a_{mk}| = \infty$, for some m, then, we can choose a strictly increasing sequence $\{k(i)\}$ of positive integers such that

$$\sum_{k=k(i)+1}^{k(i+1)} |a_{mk}| > i, \ i = 1, 2, \ldots.$$

Define the sequence $\{x_k\}$, where

$$
\begin{aligned}
x_k \quad &= \ 0, \ \text{if } 0 \leq k \leq k(1); \\
&= \ 1, \quad \text{if } a_{mk} \geq 0 \\
&= -1, \quad \text{if } a_{mk} < 0
\end{aligned}
\left.\begin{aligned}
\\
\\
\end{aligned}\right\}
\begin{aligned}
k(i) < k \leq k(i+1), \\
i = 1, 2, \ldots.
\end{aligned}
$$

Then

$$\sum_{k=0}^{\infty} a_{mk} x_k = \sum_{i=1}^{\infty} \sum_{k=k(i)+1}^{k(i+1)} |a_{mk}|$$

$$> \sum_{i=1}^{\infty} i$$

$$= \infty,$$

contradicting the fact that $\{x_k\}$ is summable A. Next, we show that

$$\sup_{n \geq 0} \sum_{k=0}^{\infty} |a_{nk}| < \infty.$$

It is clear that when A sums all sequences of 0's and 1's,

$$\lim_{n \to \infty} a_{nk} \ \text{exists}, \ k = 0, 1, 2, \ldots.$$

If possible, let

$$\sup_{n \geq 0} \sum_{k=0}^{\infty} |a_{nk}| = \infty.$$

Then we can choose two strictly increasing sequences $\{n(i)\}$, $\{k(i)\}$ of positive integers such that

$$\sum_{k=k(i)+1}^{k(i+1)} |a_{n(i),k}| > M_i + i + 1,$$

$$\sum_{k=k(i+1)+1}^{\infty} |a_{n(i),k}| < 1,$$

where $M_i = \sup_{n \geq 0} \sum_{k=0}^{k(i)} |a_{nk}|$. Define $x = \{x_k\}$ by

$$
\begin{aligned}
x_k \quad &= \quad 0, \ \text{if } 0 \leq k \leq k(1); \\
&= \quad 1, \quad \text{if } a_{n(i),k} \geq 0 \quad \left.\vphantom{\begin{array}{c}1\\1\end{array}}\right\} \quad k(i) < k \leq k(i+1), \\
&= -1, \quad \text{if } a_{n(i),k} < 0 \quad \left.\vphantom{\begin{array}{c}1\\1\end{array}}\right\} \quad i = 1, 2, \ldots.
\end{aligned}
$$

Then

$$
\sum_{k=0}^{\infty} a_{n(i),k} x_k \geq \sum_{k=k(i)+1}^{k(i+1)} |a_{n(i),k}| - \sum_{k=0}^{k(i)} |a_{n(i),k}|
$$

$$
- \sum_{k=k(i+1)+1}^{\infty} |a_{n(i),k}|
$$

$$
> M_i + i + 1 - M_i - 1
$$

$$
= i, \ i = 1, 2, \ldots,
$$

which is again a contradiction, since $\{x_k\}$ is summable A. Let, now, $b_{nk} = a_{nk} - \alpha_k$, where $\alpha_k = \lim_{n \to \infty} a_{nk}$, $k = 0, 1, 2, \ldots$. To prove that $\sum_{k=0}^{\infty} |a_{nk}|$ converges uniformly in n, it is enough to prove that $\sum_{k=0}^{\infty} |b_{nk}| \to 0$, $n \to \infty$. For, if $\sum_{k=0}^{\infty} |b_{nk}| \to 0$, $n \to \infty$, since already $\sum_{k=0}^{\infty} |b_{nk}| < \infty$, $n = 0, 1, 2, \ldots$, it follows that $\sum_{k=0}^{\infty} |b_{nk}|$ and consequently $\sum_{k=0}^{\infty} |a_{nk}|$ converges uniformly in n. If possible, let $\sum_{k=0}^{\infty} |b_{nk}| \to t$, $n \to \infty$, $t > 0$ (we assume this for convenience, though it is true for a subsequence $\{n(i)\}$ of positive integers). It is now possible to choose two strictly increasing sequences $\{n(i)\}$, $\{k(i)\}$ of positive integers such that

$$\left| \sum_{k=0}^{\infty} |b_{n(i),k}| - t \right| < \frac{t}{10},$$

$$\sum_{k=0}^{k(i)} |b_{n(i),k}| < \frac{t}{10},$$

and

$$\sum_{k=k(i+1)+1}^{\infty} |b_{n(i),k}| < \frac{t}{10},$$

so that

$$\left| \sum_{k=k(i)+1}^{k(i+1)} |b_{n(i),k}| - t \right| < \frac{3t}{10}.$$

If the sequence $x = \{x_k\}$ is defined by

$$\begin{aligned} x_k \quad &= 0, \text{ if } 0 \leq k \leq k(1); \\[2mm] &= (-1)^i, \qquad \text{if } b_{n(i),k} \geq 0 \\ &= (-1)^{i+1}, \quad \text{if } b_{n(i),k} < 0 \end{aligned} \left.\begin{aligned} \\ \\ \end{aligned}\right\}, \begin{aligned} &k(i) < k \leq k(i+1), \\[2mm] &i = 1, 2, \ldots, \end{aligned}$$

$$\begin{aligned} \left| \sum_{k=0}^{\infty} b_{n(i),k} x_k - (-1)^i t \right| \\[2mm] &\leq \frac{t}{5} + \left| \sum_{k=k(i)+1}^{k(i+1)} b_{n(i),k} x_k - (-1)^i t \right| \\[2mm] &= \frac{t}{5} + \left| (-1)^i \left\{ \sum_{k=k(i)+1}^{k(i+1)} |b_{n(i),k}| - t \right\} \right| \\[2mm] &\leq \frac{t}{5} + \frac{3t}{10} \\[2mm] &= \frac{t}{2}, \end{aligned}$$

from which it follows that $\left\{ \sum_{k=0}^{\infty} b_{nk} x_k \right\}_{n=0}^{\infty}$ is not Cauchy and so not convergent, which is a contradiction, completing the proof of the theorem. $\qquad \square$

In view of Theorem 13, the above Theorem 14 has the following improvement.

Theorem 15. *When $K = \mathbb{R}$ or $\mathbb{C}$, any matrix which sums all sequences in $\mathcal{N}P$ is necessarily a Schur matrix.*

Following Sember and Freedman [10], we now make a further study of sequences of 0's and 1's in non-Archimedean analysis (see [7], [8]). We write the sequence $\{x_k\}$, beginning with $k = 1$, for the sake of convenience.

Definition 24. *A class φ of subsets of $\mathbb{N}$, the set of all positive integers, is said to be "non-Archimedean full" if*

(i) $\displaystyle\bigcup_{S \in \varphi} S = \mathbb{N}$ *(covering);*

(ii) *if $S \subset T$ where $T \in \varphi$, then $S \in \varphi$ (hereditary);*

and

(iii) *if $\{t_k\}$ is a sequence in K such that $\displaystyle\sup_{k \in S} |t_k| < \infty$ for every $S \in \varphi$, then $\displaystyle\sup_{k \geq 1} |t_k| < \infty$.*

Example 3. *$\varphi = 2^{\mathbb{N}}$ is an example of a non-Archimedean full class.*

Theorem 16. *Let φ be a class of subsets of $\mathbb{N}$ satisfying (i) and (ii) of Definition 24. Then φ is non-Archimedean full if and only if for any infinite matrix (a_{nk}) for which*

$$\sup_{n \geq 1} \left(\sup_{k \in S} |a_{nk}| \right) < \infty,$$

for every $S \in \varphi$, then

$$\sup_{n,k} |a_{nk}| < \infty.$$

Proof. Necessity. Let φ be non-Archimedean full. Suppose, for some infinite matrix (a_{nk}), $\displaystyle\sup_{n \geq 1} \left(\sup_{k \in S} |a_{nk}| \right) < \infty$, for every $S \in \varphi$ but $\displaystyle\sup_{n,k} |a_{nk}| = \infty$. We can then choose strictly increasing sequences $\{n(j)\}$, $\{k(j)\}$ of positive integers such that

$$M(j) = \sup_{k(j-1) < i \leq k(j)} |a_{n(j),i}| > \frac{1}{\rho^{2j}},$$

where, since K is non-trivially valued, $\pi \in K$ is such that $0 < \rho = |\pi| < 1$.

Let $\mathbb{N}(j) = \{i : k(j-1) < i \leq k(j)\}$, $j = 1, 2, \ldots, k(0) = 1$

Now, define

$$b_i = a_{n(j),i}\pi^j, \ i \in \mathbb{N}(j), j = 1, 2, \ldots.$$

Then,

$$\sup_{i \in \mathbb{N}(j)} |b_i| = \sup_{i \in \mathbb{N}(j)} |a_{n(j),i}|\rho^j$$

$$= \rho^j M(j)$$

$$> \rho^j \frac{1}{\rho^{2j}}$$

$$= \frac{1}{\rho^j},$$

so that

$$\sup_{i \geq 1} |b_i| = \infty,$$

since $\frac{1}{\rho^j} \to \infty$, $j \to \infty$, $\frac{1}{\rho} > 1$. Since φ is non-Archimedean full, there exists $S \in \varphi$ with $\sup_{i \in S} |b_i| = \infty$. Consequently we have

$$\sup_{i \in S \cap \mathbb{N}(j)} |b_i| > 1 \ \text{ for infinitely many } j's,$$

for, otherwise,

$$\sup_{i \in S \cap \mathbb{N}(j)} |b_i| \leq 1, \ j = 1, 2, \ldots$$

and so

$$\sup_{i \geq 1} |b_i| \leq 1,$$

a contradiction. Hence, for these infinitely many j's,

$$\sup_{i \in S} |a_{n(j),i}| \geq \sup_{i \in S \cap \mathbb{N}(j)} |a_{n(j),i}|$$

$$= \sup_{i \in S \cap \mathbb{N}(j)} \frac{|b_i|}{\rho^j}$$

$$> \frac{1}{\rho^j} \to \infty, \ j \to \infty, \ \text{since } \frac{1}{\rho} > 1,$$

contradicting the fact that

$$\sup_{n \geq 1} \left(\sup_{k \in S} |a_{nk}| \right) < \infty, \quad \text{for every } S \in \varphi.$$

Sufficiency. Let $\{t_k\}$ be any sequence in K such that

$$\sup_{k \in S} |t_k| < \infty \quad \text{for every } S \in \varphi.$$

Define the matrix (a_{nk}), where $a_{nk} = t_k$, $k = 1, 2, \ldots$; $n = 1, 2, \ldots$. Then

$$\sup_{n \geq 1} \left(\sup_{k \in S} |a_{nk}| \right) < \infty, \quad \text{for every } S \in \varphi.$$

By hypothesis,

$$\sup_{n,k} |a_{nk}| < \infty.$$

It now follows that

$$\sup_{k \geq 1} |t_k| < \infty$$

and so φ is non-Archimedean full, completing the proof of the theorem. $\quad\square$

Corollary 1. *φ is a class of subsets of $\mathbb{N}$ satisfying (i) and (ii) of Definition 24. Then φ is non-Archimedean full if and only if for any infinite matrix (a_{nk}) for which*

$$\sup_{n \geq 1} \left| \sum_{k \in S} a_{nk} \right| < \infty, \quad \text{for every } S \in \varphi,$$

then

$$\sup_{n,k} |a_{nk}| < \infty.$$

Proof. Necessity. Let φ be non-Archimedean full. Let (a_{nk}) be an infinite matrix for which

$$\sup_{n \geq 1} \left| \sum_{k \in S} a_{nk} \right| < \infty, \quad \text{for every } S \in \varphi.$$

Let $S \in \varphi$ and $k_0 \in S$. Since φ is hereditary,

$$S' = S - \{k_0\} \in \varphi.$$

So

$$\sup_{n \geq 1} \left| \sum_{k \in S} a_{nk} - \sum_{k \in S'} a_{nk} \right| < \infty,$$

$$i.e., \ \sup_{n \geq 1} |a_{n,k_0}| < \infty,$$

for every $k_0 \in S$ and so

$$\sup_{n \geq 1} \left(\sup_{k \in S} |a_{nk}| \right) < \infty, \quad \text{for every } S \in \varphi.$$

Since φ is non-Archimedean full, it follows from Theorem 16 that

$$\sup_{n,k} |a_{n,k}| < \infty.$$

Sufficiency. Let (a_{nk}) be an infinite matrix such that

$$\sup_{n \geq 1} \left(\sum_{k \in S} |a_{nk}| \right) < \infty, \quad \text{for every } S \in \varphi.$$

Then

$$\sup_{n \geq 1} \left| \sum_{k \in S} a_{nk} \right| \leq \sup_{n \geq 1} \left(\sup_{k \in S} |a_{nk}| \right)$$

$$< \infty, \quad \text{for every } S \in \varphi.$$

By hypothesis,

$$\sup_{n,k} |a_{nk}| < \infty$$

and so φ is non-Archimedean full, using Theorem 16, completing the proof. $\square$

The following result is worth recording.

Theorem 17. *There is no minimal non-Archimedean full class.*

Proof. Let S_0 be any infinite subset of a non-Archimedean full class φ and

$$\Delta = \{S \in \varphi / S_0 \nsubseteq S\}.$$

Then $\Delta \subsetneqq \varphi$ and Δ satisfies conditions (i) and (ii) of Definition 24. Let $\{t_k\}$ be a sequence in K such that $\sup_{k \geq 1} |t_k| = \infty$. Since φ is non-Archimedean full, there exists $W \in \varphi$ such that $\sup_{k \in W} |t_k| = \infty$. So $\sup_{k \in W - S_0} |t_k| = \infty$ or $\sup_{k \in W \cap S_0} |t_k| = \infty$. In the first case, if $T = W - S_0$, then $T \in \Delta$ and $\sup_{k \in T} |t_k| = \infty$. In the second case, let $T = S_0 - \{s\}$, where $s \in S_0$. Then $T \in \Delta$ and $\sup_{k \in T} |t_k| \geq \sup_{k \in W \cap S_0} |t_k| = \infty$. In view of Definition 24, Δ is non-Archimedean full, where $\Delta \subsetneq \varphi$. Thus there is no minimal non-Archimedean full class, completing the proof. $\square$

We define

$$\chi_\varphi = \{\chi_S : S \in \varphi\},$$

where χ_S denotes the characteristic function of the subset S of $\mathbb{N}$. C_A denotes the set of all sequences $x = \{x_k\}$ which are A-summable.

As an application to matrix summability, we prove the following result.

Theorem 18. *Let φ be a non-Archimedean full class and $A = (a_{nk})$ be any infinite matrix. Then $C_A \supseteq \chi_\varphi$ if and only if*

(i) $\lim_{k \to \infty} a_{nk} = 0$, $n = 1, 2, \ldots$;

   and

(ii) $\lim_{n \to \infty} \sup_{k \in S} |a_{n+1,k} - a_{nk}| = 0$ *for every $S \in \varphi$.*

Proof. Necessity. Let $C_A \supseteq \chi_\varphi$. It is clear that (i) holds. So

$$\lim_{k \to \infty} (a_{n+1,k} - a_{nk}) = 0.$$

Suppose (ii) does not hold. We use the "sliding hump method" (or the Ganapathy Iyer-Schur method) to reach a contradiction. We can now choose $\epsilon > 0$, $S \in \varphi$ and two strictly increasing sequences $\{n(i)\}$ and $\{k(i)\}$ of positive integers such that

$$\sup_{k \in S} |a_{n(i)+1,k} - a_{n(i),k}| > \epsilon;$$

$$\sup_{1 \leq k \leq k(i-1)} |a_{n(i)+1,k} - a_{n(i),k}| < \frac{\epsilon}{8};$$

and

$$\sup_{k > k(i)} |a_{n(i)+1,k} - a_{n(i),k}| < \frac{\epsilon}{8}.$$

In view of the above inequalities, there exists $k(n(i)) \in S$, $k(i-1) < k(n(i)) \leq k(i)$ such that

$$|a_{n(i)+1,k(n(i))} - a_{n(i),k(n(i))}| > \epsilon.$$

Define $x = \{x_k\}$, where

$$x_k = 1, \text{ if } k = k(n(i));$$

$$= 0, \text{ otherwise.}$$

Now,

$$(Ax)_{n(i)+1} - (Ax)_{n(i)}$$

$$= \sum_{k=1}^{\infty} \{a_{n(i)+1,k} - a_{n(i),k}\} x_k$$

$$= \sum_{k=1}^{k(i-1)} \{a_{n(i)+1,k} - a_{n(i),k}\} x_k$$

$$+ \sum_{k=k(i-1)+1}^{k(i)} \{a_{n(i)+1,k} - a_{n(i),k}\} x_k$$

$$+ \sum_{k=k(i)+1}^{\infty} \{a_{n(i)+1,k} - a_{n(i),k}\} x_k$$

$$= \sum_{k=1}^{k(i-1)} \{a_{n(i)+1,k} - a_{n(i),k}\} x_k$$

$$+ \{a_{n(i)+1,k(n(i))} - a_{n(i),k(n(i))}\}$$

$$+ \sum_{k=k(i)+1}^{\infty} \{a_{n(i)+1,k} - a_{n(i),k}\} x_k,$$

so that

$$\epsilon < |a_{n(i)+1,k(n(i))} - a_{n(i),k(n(i))}|$$

$$\leq \max \left[|(Ax)_{n(i)+1} - (Ax)_{n(i)}|, \frac{\epsilon}{8}, \frac{\epsilon}{8} \right],$$

which implies that

$$|(Ax)_{n(i)+1} - (Ax)_{n(i)}| > \epsilon, \ i = 1, 2, \ldots.$$

Thus $x \notin C_A$. Note, however, that $x \in \chi_\varphi$. Consequently, $\chi_\varphi \nsubseteq C_A$, a contradiction. So (ii) holds.

Sufficiency. Let (i), (ii) hold. In view of (i), $\sum_{k \in S} a_{nk}$ converges for every $S \in \varphi$.

Now,

$$\left| \sum_{k \in S} a_{n+1,k} - \sum_{k \in S} a_{nk} \right| = \left| \sum_{k \in S} (a_{n+1,k} - a_{nk}) \right|$$

$$\leq \sup_{k \in S} |a_{n+1,k} - a_{nk}|$$

$$\to 0, n \to \infty, \text{ using } (ii),$$

which implies that $\lim_{n \to \infty} \sum_{k \in S} a_{nk}$ exists for every $S \in \varphi$. Thus $C_A \supseteq \chi_\varphi$, completing the proof of the theorem. □

Corollary 2 (Hahn's theorem for the non-Archimedean case). *An infinite matrix $A = (a_{nk})$ sums all bounded sequences if and only if it sums all sequences of 0's and 1's.*

Proof. Leaving out the trivial part of the result, suppose A sums all sequences of 0's and 1's, i.e., $C_A \supseteq \chi_\varphi$, where $\varphi = 2^{\mathbb{N}}$. Since $\mathbb{N} \in \varphi$,

$$\lim_{k \to \infty} a_{nk} = 0, n = 1, 2, \ldots.$$

Also,

$$\lim_{n \to \infty} \sup_{k \in \mathbb{N}} |a_{n+1,k} - a_{nk}| = 0,$$

$$i.e., \ \lim_{n \to \infty} \sup_{k \geq 1} |a_{n+1,k} - a_{nk}| = 0.$$

In view of Theorem 2 of [3], A sums all bounded sequences, completing the proof. □

In the context of the space Λ_r, it is of relevance to introduce the sequence space Q as in the following:

Definition 25. *The sequence* $x = \{x_k\}$ *belongs to* Q, *if for any* $\epsilon > 0$, *there exist positive integers* n, k_0 *such that*

$$|x_k - x_{k+sn}| < \epsilon, \ \ k \geq k_0, s = 0, 1, 2, \dots.$$

Note that Q *is a linear subspace of* ℓ_∞. *It is further closed, when* K *is complete. For, let* $\{x^{(n)}\}$ *be a Cauchy sequence in* Q, *where* $x^{(n)} = \{x_k^{(n)}\}_{k=0}^\infty$. *So, given any* $\epsilon > 0$, *there exists a positive integer* n_0 *such that*

$$\|x^{(n)} - x^{(m)}\| < \frac{\epsilon}{3}, \ \ n, m \geq n_0,$$

$$i.e., \ \sup_{k \geq 0} |x_k^{(n)} - x_k^{(m)}| < \frac{\epsilon}{3}, \ \ n, m \geq n_0. \tag{2.3}$$

So, for each $k = 0, 1, 2, \dots, \{x_k^{(n)}\}_{n=0}^\infty$ *is a Cauchy sequence in the complete field* K *and hence there exists* $x_k \in K$ *such that*

$$x_k^{(n)} \to x_k, \ \ n \to \infty, k = 0, 1, 2, \dots.$$

Let $x = \{x_k\}$. *We now claim that*

$$x^{(n)} \to x, \ \ n \to \infty$$

in ℓ_∞. *To prove this, we note that, since* $x^{(n_0)} \in Q$, *given* $\epsilon > 0$, *there exist positive integers* u, k_0 *such that*

$$|x_k^{(n_0)} - x_{k+su}^{(n_0)}| < \frac{\epsilon}{3}, \ \ k \geq k_0, s = 0, 1, 2, \dots.$$

In (2.3), for $n \geq n_0$, *letting* $m \to \infty$,

$$\sup_{k \geq 0} |x_k^{(n)} - x_k| \leq \frac{\epsilon}{3}, \ \ n \geq n_0,$$

$$i.e., \|x^{(n)} - x\| \leq \frac{\epsilon}{3}, \ \ n \geq n_0,$$

which shows that

$$x^{(n)} \to x, \ \ n \to \infty$$

in ℓ_∞. Now,

$$|x_k - x_{k+su}| \leq |x_k - x_k^{(n_0)}| + |x_k^{(n_0)} - x_{k+su}^{(n_0)}|$$
$$+ |x_{k+su}^{(n_0)} - x_{k+su}|$$
$$< \frac{\epsilon}{3} + \frac{\epsilon}{3} + \frac{\epsilon}{3}$$
$$= \epsilon, \quad k \geq k_0, s = 0, 1, 2, \ldots,$$

thus proving that $x \in Q$. Consequently, Q is complete and so a closed subspace of ℓ_∞.

It is easy to prove that Λ_r is a closed subspace of ℓ_∞. We leave the details to the reader. Further,

$$Q \subset \overline{\left(\bigcup_{r=1}^{\infty} \Lambda_r \right)},$$

where the right-hand side denotes the closure of $\bigcup_{r=1}^{\infty} \Lambda_r$. For, if $x = \{x_k\} \in Q$, given $\epsilon > 0$, there exist positive integers n, k_0 such that

$$|x_k - x_{k+sn}| < \epsilon, \quad k \geq k_0, s = 0, 1, 2, \ldots.$$

The sequence $y = \{y_k\}$, where

$$
\begin{aligned}
y_k \ &= x_k, && \text{if } 0 \leq k \leq k_0 + n - 1; \\
&= x_{k_0+j}, && \text{if } k_0 + (i+1)n \leq k < k_0 + (i+2)n, \\
& && k = k_0 + (i+1)n + j, \\
& && j = 0, 1, 2, \ldots, (n-1), \\
& && i = 0, 1, 2, \ldots,
\end{aligned}
$$

is in Λ_n and $\|x - y\| < \epsilon$. However, when K is a complete, non-trivially valued, non-Archimedean field, we have, in fact,

$$Q = \overline{\left(\bigcup_{r=1}^{\infty} \Lambda_r \right)}.$$

To establish this fact, it remains to prove that

$$\overline{\left(\bigcup_{r=1}^{\infty} \Lambda_r \right)} \subset Q.$$

To see this, if $x = \{x_k\} \in \overline{\left(\bigcup_{r=1}^{\infty} \Lambda_r \right)}$, there exists a positive integer n and $y = \{y_k\} \in \Lambda_n$ such that $\|x - y\| < \epsilon$. This implies

$$|x_k - x_{k+sn}| \leq \max\{|x_{k+sn} - y_{k+sn}|, |y_{k+sn} - y_k|,$$

$$|y_k - x_k|\}$$

$$< \epsilon, \ k \geq k_0, s = 0, 1, 2, \ldots,$$

$$i.e., x \in Q,$$

completing the proof.

2.4 The Steinhaus theorem

Somasundaram [11] obtained the Steinhaus theorem (Theorem 21) over complete, non-trivially valued, non-Archimedean fields for a restricted class of regular matrices and noted that the theorem was not true in general. His restriction is, in fact, meaningless and his observation is incorrect. Neither of the examples given in Section 2 of [11] provides a regular matrix which sums all bounded sequences.

The following theorem generalizes the characterization of Schur matrices in the non-Archimedean case and establishes, in particular, the validity of the Steinhaus theorem in this case too.

Theorem 19. $A = (a_{nk})$, $a_{nk} \in K$, $n, k = 0, 1, 2, \ldots$ *is an infinite matrix.*

(i) If $K = \mathbb{R}$ or $\mathbb{C}$, $A \in (\ell_\infty, \Lambda_r)$ if and only if

$$\left. \begin{array}{c} (a) \sup_{n \geq 0} \sum_{k=0}^{\infty} |a_{nk}| < \infty; \\[2mm] and \\[2mm] (b) \lim_{n \to \infty} \sum_{k=0}^{\infty} |a_{n+r,k} - a_{nk}| = 0. \end{array} \right\} \qquad (2.4)$$

(ii) (see [3]) If K is a complete, non-trivially valued, non-Archimedean field, $A \in (\ell_\infty, \Lambda_r)$ if and only if

$$(a) \; \lim_{k \to \infty} a_{nk} = 0, \; n = 0, 1, 2, \dots;$$

$$and$$

$$(b) \; \lim_{n \to \infty} \sup_{k \geq 0} |a_{n+r,k} - a_{nk}| = 0. \tag{2.5}$$

Proof. (i) Let $K = \mathbb{R}$ or $\mathbb{C}$. Sufficiency. Suppose (2.4) holds. Let $x = \{x_k\} \in \ell_\infty$. In view of (2.4)(a), $\sum_{k=0}^{\infty} a_{nk} x_k$ converges, $n = 0, 1, 2, \dots$. Again, by (2.4)(a), $\{(Ax)_n\} \in \ell_\infty$. Also,

$$|(Ax)_{n+r} - (Ax)_n| = \left| \sum_{k=0}^{\infty} (a_{n+r,k} - a_{nk}) x_k \right|$$

$$\leq M \sum_{k=0}^{\infty} |a_{n+r,k} - a_{nk}|$$

$$\to 0, \; n \to \infty,$$

in view of (2.4)(b), where $M = \sup_{k \geq 0} |x_k|$. Thus $\{(Ax)_n\} \in \Lambda_r$.

Necessity. It is clear that (2.4)(a) is necessary. Suppose (2.4)(b) does not hold. Then, for $\epsilon > 0$, there exist sequences $\{n(m)\}, \{k(m)\}$ of positive integers such that $n(m) > n(m-1)$, $k(m) > k(m-1)$, with

$$\sum_{k=0}^{k(m-1)} |a_{n(m)+r,k} - a_{n(m),k}| < \frac{\epsilon}{4}; \tag{2.6}$$

$$\sum_{k=0}^{\infty} |a_{n(m)+r,k} - a_{n(m),k}| > \epsilon; \tag{2.7}$$

and

$$\sum_{k=k(m)+1}^{\infty} |a_{n(m)+r,k} - a_{n(m),k}| < \frac{\epsilon}{8}. \tag{2.8}$$

Define the sequence $x = \{x_k\}$ by

$$x_k \; = 0, \text{ if } 0 \leq k \leq k(1);$$

$$= sgn\{a_{n(m)+r,k} - a_{n(m),k}\},$$

$$\text{if } k(m-1) < k \leq k(m), m = 2, 3, \dots.$$

Using (2.6), (2.7) and (2.8), we have,

$$\sum_{k=k(m-1)+1}^{k(m)} |a_{n(m)+r,k} - a_{n(m),k}| > \epsilon - \frac{\epsilon}{4} - \frac{\epsilon}{8}$$

$$> \frac{\epsilon}{2},$$

so that

$$|(Ax)_{n(m)+r} - (Ax)_{n(m)}|$$

$$\geq \sum_{k=k(m-1)+1}^{k(m)} |a_{n(m)+r,k} - a_{n(m),k}|$$

$$- \sum_{k=0}^{k(m-1)} |a_{n(m)+r,k} - a_{n(m),k}|$$

$$- \sum_{k=k(m)+1}^{\infty} |a_{n(m)+r,k} - a_{n(m),k}|$$

$$> \frac{\epsilon}{2} - \frac{\epsilon}{4} - \frac{\epsilon}{8}$$

$$= \frac{\epsilon}{8},$$

this being true for $m = 2, 3, \ldots$. Noting that $x = \{x_k\} \in \ell_\infty$, while $\{(Ax)_n\} \notin \Lambda_r$, we have a contradiction. Thus (2.4)(b) is necessary, completing the proof.

(ii) Let K be a complete, non-trivially valued, non-Archimedean field.

Sufficiency. First note that (2.5) implies that

$$\sup_{n,k} |a_{nk}| < \infty.$$

For, by (2.5)(b), given $\epsilon > 0$, there exists for $k = 0, 1, 2, \ldots, n_0 = n_0(\epsilon)$, such that

$$|a_{n+r,k} - a_{nk}| < \epsilon, \; n \geq n_0.$$

Without loss of generality, let us suppose that $n_0 \equiv 0 \pmod{r}$. If $n \equiv i \pmod{r}$, $i = 0, 1, 2, \ldots, (r-1)$, then,

$$|a_{nk} - a_{n_0+i,k}| \leq \max\{|a_{nk} - a_{n-r,k}|, |a_{n-r,k} - a_{n-2r,k}|,$$

$$\ldots, |a_{n_0+i+r,k} - a_{n_0+i,k}|\}$$

$$< \epsilon, \; n \geq n_0 + i.$$

Hence if $n \geq n_0 + i$,

$$|a_{nk}| \leq \max\{\epsilon, |a_{n_0+i,k}|\}$$

$$\leq \max\{\epsilon, M(n_0 + i)\},$$

where $M(n_0 + i) > 0$ is such that $|a_{n_0+i,k}| \leq M(n_0 + i)$, $k = 0, 1, 2, \ldots$, this being so because of (2.5)(a). If $M_1 = \max \ M(n_0 + i)$, $i = 0, 1, 2, \ldots, (r - 1)$, then

$$|a_{nk}| \leq \max\{\epsilon, M_1\}, \ n \geq n_0, k = 0, 1, 2, \ldots.$$

Again, by (2.5)(a), there exists $M_2 > 0$ with

$$|a_{nk}| \leq M_2, \ n = 0, 1, 2, \ldots, (n_0 - 1); k = 0, 1, 2, \ldots.$$

Consequently,

$$|a_{nk}| < \max\{M_1 + \epsilon, M_2\}, \ n, k = 0, 1, 2, \ldots,$$

proving that

$$\sup_{n,k} |a_{nk}| < \infty.$$

Let (2.5) hold and $x = \{x_k\} \in \ell_\infty$. Noting that $a_{nk}x_k \to 0$, $k \to \infty$, it follows that $\sum_{k=0}^{\infty} a_{nk}x_k$ converges, $n = 0, 1, 2, \ldots$. Further,

$$|(Ax)_{n+r} - (Ax)_n| = \left| \sum_{k=0}^{\infty} (a_{n+r,k} - a_{nk})x_k \right|$$

$$\leq M \sup_{k \geq 0} |a_{n+r,k} - a_{nk}|$$

$$\to 0, \ n \to \infty,$$

where $M = \sup_{k \geq 0} |x_k|$. Also,

$$\sup_{n \geq 0} |(Ax)_n| \leq M \sup_{n,k} |a_{nk}| < \infty,$$

from which it follows that $\{(Ax)_n\} \in \ell_\infty$. Consequently, $\{(Ax)_n\} \in \Lambda_r$.

Necessity. Let $A \in (\ell_\infty, \Lambda_r)$. We note that $(Ax)_n = \sum_{k=0}^{\infty} a_{nk}$ if $x_k = 1$, $k = $

$0, 1, 2, \ldots$ and that $(Ax)_n$ will be defined for each $n = 0, 1, 2, \ldots$, only if (2.5)(a) holds. Consider the sequence $\{x_n\}$, $x_n = 0$, $n \neq k$, $x_k = 1$, for which $(Ax)_n = a_{nk}$, $n = 0, 1, 2, \ldots$, we have from $\{(Ax)_n\} \in \Lambda_r$,

$$|a_{n+r,k} - a_{nk}| \to 0, \ n \to \infty, k = 0, 1, 2, \ldots .$$

If possible, let (2.5)(b) be false. Then, for some $\epsilon > 0$, we have increasing sequences $\{n(i)\}$, $\{k(n(i))\}$, $\{k(i)\}$ of positive integers, with $k(n(i)) > k(i) \geq k(n(i-1))$ such that

$$\sup_{0 \leq k < k(n(i-1))} |a_{n(i)+r,k} - a_{n(i),k}| < \frac{\epsilon}{2};$$

$$\sup_{k(n(i)) \leq k < \infty} |a_{n(i)+r,k} - a_{n(i),k}| < \frac{\epsilon}{2};$$

and

$$|a_{n(i)+r,k(i)} - a_{n(i),k(i)}| > \epsilon, \ i = 1, 2, \ldots . \tag{2.9}$$

Define the sequence $x = \{x_k\} \in \ell_\infty$ by

$$\left. \begin{array}{l} x_k &= 1, \ k = k(i) \\ &= 0, \ k \neq k(i) \end{array} \right\}, \ i = 1, 2, \ldots .$$

Then,

$$(Ax)_{n(i)+r} - (Ax)_{n(i)}$$

$$= \sum_{k=0}^{k(n(i-1))-1} \{a_{n(i)+r,k} - a_{n(i),k}\} x_k$$

$$+ \sum_{k=k(n(i))}^{\infty} \{a_{n(i)+r,k} - a_{n(i),k}\} x_k$$

$$+ \sum_{k=k(n(i-1))}^{k(n(i))-1} \{a_{n(i)+r,k} - a_{n(i),k}\} x_k$$

$$= \sum_{k=0}^{k(n(i-1))-1} \{a_{n(i)+r,k} - a_{n(i),k}\} x_k$$

$$+ \sum_{k=k(n(i))}^{\infty} \{a_{n(i)+r,k} - a_{n(i),k}\} x_k$$

$$+ \{a_{n(i)+r,k(i)} - a_{n(i),k(i)}\}.$$

Thus,

$$|a_{n(i)+r,k(i)} - a_{n(i),k(i)}|$$

$$\leq \max\{|(Ax)_{n(i)+r} - (Ax)_{n(i)}|,$$

$$\sup_{0 \leq k < k(n(i-1))} |a_{n(i)+r,k} - a_{n(i),k}|,$$

$$\sup_{k(n(i)) \leq k < \infty} |a_{n(i)+r,k} - a_{n(i),k}|\}. \tag{2.10}$$

In view of (2.9) and (2.10), we have,

$$\epsilon < |a_{n(i)+r,k(i)} - a_{n(i),k(i)}|$$

$$\leq \max\{|(Ax)_{n(i)+r} - (Ax)_{n(i)}|, \frac{\epsilon}{2}, \frac{\epsilon}{2}\},$$

so that

$$|(Ax)_{n(i)+r} - (Ax)_{n(i)}| > \epsilon, \ i = 2, 3, \ldots.$$

Hence,

$$(Ax)_{n+r} - (Ax)_n \not\to 0, \ n \to \infty,$$

$$i.e., \ \{(Ax)_n\} \notin \Lambda_r,$$

while, $x = \{x_k\} \in \ell_\infty$, which is a contradiction. Thus (2.5)(b) is necessary, completing the proof. $\square$

Remark 2. *Theorem 19(i) can also be deduced from the characterization of infinite matrices belonging to the matrix class (ℓ_∞, c_0). It is known [12] that $A \in (\ell_\infty, c_0)$ if and only*

$$\sum_{k=0}^{\infty} |a_{nk}| \to 0, \ n \to \infty.$$

Clearly (2.4)(a) and (b) are sufficient for $A \in (\ell_\infty, \Lambda_r)$. It is enough to prove the necessity of (2.4)(b). Let

$$b_{nk} = a_{n+r,k} - a_{nk}, \ n, k = 0, 1, 2, \ldots.$$

Since $A \in (\ell_\infty, \Lambda_r)$, if $x = \{x_k\} \in \ell_\infty$, then $\{(Ax)_n\} \in \Lambda_r$. Thus,

$$\sum_{k=0}^{\infty} b_{nk} x_k = \sum_{k=0}^{\infty} (a_{n+r,k} - a_{nk}) x_k$$

$$\to 0, \ n \to \infty.$$

Hence the infinite matrix $B = (b_{nk}) \in (\ell_\infty, c_0)$ and so

$$\sum_{k=0}^{\infty} |b_{nk}| \to 0, \ n \to \infty,$$

i.e., $\displaystyle \sum_{k=0}^{\infty} |a_{n+r,k} - a_{nk}| \to 0, \ n \to \infty,$

so that (2.4)(b) is necessary.

If K is a complete, non-trivially valued, non-Archimedean field, $\Lambda_1 = c$ and so we have the characterization of Schur matrices in the non-Archimedean case.

Theorem 20. *A is a Schur matrix, i.e., $A \in (\ell_\infty, c)$ if and only if*

$$(a) \ \lim_{k \to \infty} a_{nk} = 0, \ n = 0, 1, 2, \ldots;$$

and

$$(b) \ \lim_{n \to \infty} \sup_{k \geq 0} |a_{n+1,k} - a_{nk}| = 0.$$

Remark 3. *It is implicit in the proof of Theorem 19(ii), for the case $r = 1$, that if A sums all sequences of 0's and 1's, then A is a Schur matrix in the non-Archimedean case too (cf. Theorem 14). Since, by Theorem 13, any sequence of 0's and 1's in $\bigcup_{r=1}^{\infty} \Lambda_r$ can be expressed as the difference of two non-periodic sequences of 0's and 1's, it follows that if A sums all sequences in $\mathcal{N}P$, then A is a Schur matrix in this case also (cf. Theorem 11).*

As a consequence of Theorem 20, we have the Steinhaus theorem in the non-Archimedean case (cf. Theorem 11).

Theorem 21. *A matrix cannot be both a Toeplitz (or regular) and a Schur matrix, or, equivalently, given any regular matrix A, there exists a bounded, divergent sequence which is not summable A.*

Proof. Suppose A is both a regular and a Schur matrix. Then from the criterion of regularity of A (see [2], [6]),

$$\lim_{n \to \infty} a_{nk} = 0, \ k = 0, 1, 2, \ldots;$$

and

$$\lim_{n \to \infty} \sum_{k=0}^{\infty} a_{nk} = 1.$$

Since A is a Schur matrix too, from Theorem 20(b), it follows that $\{a_{nk}\}_{n=0}^{\infty}$ is uniformly Cauchy with respect to $k = 0, 1, 2, \ldots$ and hence converges uniformly to the limit 0 as $n \to \infty$, since K is complete. In other words,

$$\sup_{k \geq 0} |a_{nk}| \to 0, \ n \to \infty,$$

so that

$$\left| \sum_{k=0}^{\infty} a_{nk} \right| \leq \sup_{k \geq 0} |a_{nk}| \to 0, \ n \to \infty,$$

$$i.e., \ \lim_{n \to \infty} \sum_{k=0}^{\infty} a_{nk} = 0,$$

which is a contradiction. This establishes the theorem. $\qquad \square$

When $K = \mathbb{R}$ or $\mathbb{C}$, the following sets of necessary and sufficient conditions for a matrix to be a Schur matrix are known and are listed in [12] as the conditions known so far.

$$\left.\begin{array}{l} (a) \ \lim_{n \to \infty} a_{nk} \text{ exists}, \ k = 0, 1, 2, \ldots; \\[2ex] (b) \ \lim_{n \to \infty} \sum_{k=0}^{\infty} |a_{nk}| = \sum_{k=0}^{\infty} \left| \lim_{n \to \infty} a_{nk} \right|. \end{array}\right\} \qquad (2.11)$$

$$\left.\begin{array}{l} (a) \ \sup_{n \geq 0} \sum_{k=0}^{\infty} |a_{nk}| < \infty; \\[2ex] (b) \ \lim_{n \to \infty} a_{nk} \text{ exists}, \ k = 0, 1, 2, \ldots; \\[2ex] (c) \ \lim_{n \to \infty} \sum_{k=0}^{\infty} \left| a_{nk} - \lim_{n \to \infty} a_{nk} \right| = 0. \end{array}\right\} \qquad (2.12)$$

$$\left.\begin{array}{l} (a) \ \lim_{n\to\infty} a_{nk} \text{ exists, } k = 0, 1, 2, \ldots ; \\[2mm] (b) \ \sum_{k=0}^{\infty} |a_{nk}| \text{ converges uniformly in } n. \end{array}\right\} \tag{2.13}$$

In Theorem 22 below, adapting the proof of Theorem 19, we obtain a set of necessary and sufficient conditions, which seems to be different from (2.11), (2.12), (2.13), for a matrix to be a Schur matrix.

Theorem 22. *When $K = \mathbb{R}$ or $\mathbb{C}$, $A \in (\ell_\infty, c)$ if and only if*

$$\left.\begin{array}{l} (a) \ \sum_{k=0}^{\infty} |a_{nk}| < \infty, \ n = 0, 1, 2, \ldots ; \\[2mm] (b) \ \sup_{p\geq 0} \sum_{k=0}^{\infty} |a_{n+p,k} - a_{nk}| \to 0, \ n \to \infty. \end{array}\right\} \tag{2.14}$$

The reader can prove the equivalence of (2.12) and (2.14) without any difficulty.

Remark 4. *Though conditions (2.13)(b) and (2.14)(b) appear to be equivalent, they are really independent of each other as is illustrated by the following examples.*

Example 4. *For the matrix*

$$A \equiv (a_{nk}) = \begin{bmatrix} 1 + \frac{1}{1^2} & \frac{1}{2} + \frac{1}{2^2} & \frac{1}{3} + \frac{1}{3^2} & \cdots \\[2mm] 1 + \frac{1}{2}\cdot\frac{1}{1^2} & \frac{1}{2} + \frac{1}{2}\cdot\frac{1}{2^2} & \frac{1}{3} + \frac{1}{2}\cdot\frac{1}{3^2} & \cdots \\[2mm] 1 + \frac{1}{3}\cdot\frac{1}{1^2} & \frac{1}{2} + \frac{1}{3}\cdot\frac{1}{2^2} & \frac{1}{3} + \frac{1}{3}\cdot\frac{1}{3^2} & \cdots \\[2mm] \cdots & \cdots & \cdots & \cdots \end{bmatrix},$$

i.e., $a_{nk} = \dfrac{(n+1)(k+1) + 1}{(n+1)(k+1)^2}$, $n, k = 0, 1, 2, \ldots,$

(2.14)(b) is satisfied, while (2.13)(b) fails to hold.

Example 5. *For the matrix*

$$
A \equiv (a_{nk}) =
\begin{bmatrix}
0 & 1 & \frac{1}{2^2} & \frac{1}{3^2} & \cdots \\
1 & \frac{1}{2} & \frac{1}{2} \cdot \frac{1}{2^2} & \frac{1}{2} \cdot \frac{1}{3^2} & \cdots \\
2 & \frac{1}{3} & \frac{1}{3} \cdot \frac{1}{2^2} & \frac{1}{3} \cdot \frac{1}{3^2} & \cdots \\
\cdots & \cdots & \cdots & \cdots & \cdots
\end{bmatrix},
$$

i.e.,
$$
\left.
\begin{aligned}
a_{nk} &= \frac{1}{(n+1)k^2}, \ k \geq 1 \\
a_{n0} &= n
\end{aligned}
\right\}, \ n = 0, 1, 2, \ldots,
$$

(2.13)(b) holds, while (2.14)(b) is not satisfied.

Remark 5. *It is simpler to check (2.14)(a)-(b) than (2.13)(a)-(b). For instance, consider the matrix defined by*

$$
a_{nk} = \left\{ \frac{(-1)^k 2\theta \sin(k+1)\omega}{(k+1)\omega} \right\} e^{-(k+1)\lambda^2(n+1)},
$$

$n, k = 0, 1, 2, \ldots$, where θ, ω, λ are positive parameters, then, $A \equiv (a_{nk})$ is seen to be a Schur matrix on an easy checking of (2.14)(a)-(b). Checking of (2.13)(a)-(b), however, involves using Abel's test for uniform convergence.

Theorem 23. *When K is a complete, non-trivially valued, non-Archimedean field, $A \in (\ell_\infty, c_0)$ if and only if*

$$
(a) \lim_{k \to \infty} a_{nk} = 0, \ n = 0, 1, 2, \ldots;
$$

and

$$
(b) \sup_{k \geq 0} |a_{nk}| \to 0, \ n \to \infty.
$$

Proof. It is clear that conditions (a)-(b) are sufficient for $A \in (\ell_\infty, c_0)$. If $A \in (\ell_\infty, c_0)$, $(Ax)_n = \sum_{k=0}^{\infty} a_{nk}$ if $x_k = 1$, $k = 0, 1, 2, \ldots$ and that $(Ax)_n$ will be defined for each $n = 0, 1, 2, \ldots$, only if (a) holds. Considering the sequence $\{x_n\}$, $x_n = 0$, $n \neq k$, $x_k = 1$, for which $(Ax)_n = a_{nk}$, $n = 0, 1, 2, \ldots$, we have,

$$
a_{nk} \to 0, \ n \to \infty, k = 0, 1, 2, \ldots.
$$

Since A is a Schur matrix,

$$a_{nk} \to 0, n \to \infty, \quad \text{uniformly with respect to } k = 0, 1, 2, \ldots,$$

$$\text{i.e., } \sup_{k \geq 0} |a_{nk}| \to 0, \ n \to \infty,$$

which proves the necessity of (b), completing the proof. $\square$

2.5   A Steinhaus-type theorem

With the notation introduced in Section 1.6, the Steinhaus theorem can be written as

$$(c, c; P) \cap (\ell_\infty, c) = \phi.$$

Theorems of the above type are called "Steinhaus-type theorems." One such result, improving the Steinhaus theorem, is proved below.

Theorem 24 (see [4]). *When $K = \mathbb{R}$ or $\mathbb{C}$,*

$$(c, c; P) \cap \left(\Lambda_r - \bigcup_{i=1}^{r-1} \Lambda_i, c \right) = \phi.$$

Proof. We can choose two sequences $\{n(m)\}, \{k(m)\}$ of positive integers (cf. proof of Theorem 11) such that

if $m = 2p$, $n(m) > n(m-1)$, $k(m) > k(m-1) + (2m-5)r$,

$$\sum_{k=0}^{k(m-1)+(2m-5)r} |a_{n(m),k}| < \frac{1}{16};$$

$$\sum_{k=k(m)+1}^{\infty} |a_{n(m),k}| < \frac{1}{16};$$

if $m = 2p + 1$, $n(m) > n(m-1)$, $k(m) > k(m-1) + (m-2)r$,

$$\sum_{k=0}^{k(m-1)+(m-2)r} |a_{n(m),k}| < \frac{1}{16};$$

$$\left| \sum_{k=k(m-1)+(m-2)r+1}^{k(m)} a_{n(m),k} \right| > \frac{7}{8};$$

$$\sum_{k=k(m)+1}^{\infty} |a_{n(m),k}| < \frac{1}{16}.$$

Define the sequence $x = \{x_k\}$ as follows:

if $k(2p-1) < k \leq k(2p)$,

$$x_k = \begin{cases} \frac{2p-2}{2p-1}, & \text{if } k = k(2p-1) + 1; \\[6pt] 1, & \text{if } k(2p-1) + 1 < k \leq k(2p-1) + r; \\[6pt] \frac{2p-3}{2p-1}, & \text{if } k = k(2p-1) + r + 1; \\[6pt] 1, & \text{if } k(2p-1) + r + 1 < k \leq k(2p-1) + 2r; \\[6pt] \vdots \\[6pt] \frac{1}{2p-1}, & \text{if } k = k(2p-1) + (2p-3)r + 1; \\[6pt] \frac{2p-2}{2p-1}, & \text{if } k(2p-1) + (2p-3)r + 1 < k \leq k(2p-1) + (2p-2)r; \\[6pt] 0, & \text{if } k = k(2p-1) + (2p-2)r + 1; \\[6pt] \frac{2p-3}{2p-1}, & \text{if } k(2p-1) + (2p-2)r + 1 < k \leq k(2p-1) + (2p-1)r; \\[6pt] 0, & \text{if } k = k(2p-1) + (2p-1)r + 1; \\[6pt] \vdots \\[6pt] \frac{1}{2p-1}, & \text{if } k(2p-1) + (4p-6)r + 1 < k \leq k(2p-1) + (4p-5)r; \\[6pt] 0, & \text{if } k(2p-1) + (4p-5)r + 1 < k \leq k(2p); \end{cases}$$

if $k(2p) < k \le k(2p+1)$,

$$x_k = \begin{cases} \frac{1}{2p}, & \text{if } k(2p) < k \le k(2p) + r; \\[2mm] \frac{2}{2p}, & \text{if } k(2p) + r < k \le k(2p) + 2r; \\[2mm] \vdots \\[2mm] \frac{2p-1}{2p}, & \text{if } k(2p) + (2p-2)r < k \le k(2p) + (2p-1)r; \\[2mm] 1, & \text{if } k(2p) + (2p-1)r < k \le k(2p+1). \end{cases}$$

We note that, if $k(2p-1) < k \le k(2p)$,

$$|x_{k+r} - x_k| < \frac{1}{2p-1},$$

while, if $k(2p) < k \le k(2p+1)$,

$$|x_{k+r} - x_k| < \frac{1}{2p}.$$

Thus $|x_{k+r} - x_k| \to 0$, $p \to \infty$, showing that $x = \{x_k\} \in \Lambda_r$. However,

$$|x_{k+1} - x_k| = \frac{2p-2}{2p-1}, \text{ if } k = k(2p-1) + (2p-3)r, p = 1, 2, \ldots.$$

Hence, $|x_{k+1} - x_k| \nrightarrow 0$, $k \to \infty$ and consequently $x \notin \Lambda_1$. In a similar manner, we can show that $x \notin \Lambda_i$, $i = 2, 3, \ldots, (r-1)$. Thus,

$$x \in \Lambda_r - \bigcup_{i=1}^{r-1} \Lambda_i.$$

Further,

$$\left. \begin{array}{l} |(Ax)_{n(2p)}| < \frac{1}{16} + \frac{1}{16} = \frac{1}{8}, \\[2mm] |(Ax)_{n(2p+1)}| > \frac{7}{8} - \frac{1}{16} - \frac{1}{16} = \frac{3}{4} \end{array} \right\}, p = 1, 2, \ldots,$$

which shows that $\{(Ax)_n\} \notin c$, completing the proof of the theorem. $\qquad \square$

Remark 6. *We note that when* $K = \mathbb{R}$ *or* $\mathbb{C}$,

$$(c, c; P) \cap (\Lambda_r, c) = \phi. \tag{2.15}$$

A consequence of (2.15) is of some interest. Theorem 15 asserts that a matrix which sums all sequences in $\mathcal{N}P$, i.e., all non-periodic sequences of 0's and 1's, sums all bounded sequences. It is, therefore, natural to enquire whether a matrix which sums all sequences of 0's and 1's in $\bigcup_{r=1}^{\infty} \Lambda_r$ sums all sequences in $\bigcup_{r=1}^{\infty} \Lambda_r$. The answer to this query is, however, in the negative. For, by Theorem 12 and (2.15), there exists even a regular matrix which sums all sequences of 0's and 1's in $\bigcup_{r=1}^{\infty} \Lambda_r$ but fails to sum at least one sequence in $\bigcup_{r=1}^{\infty} \Lambda_r$.

In the context of the Steinhaus-type theorem (2.15), Theorem 25 below indicates a deviation from the classical case in the non-Archimedean case.

Theorem 25. *There exists a regular matrix, with entries in $\mathbb{Q}_p$, the p-adic field for a prime p, which sums all sequences in Λ_r, for a fixed positive integer r.*

Proof. Consider the matrix $A \equiv (a_{nk})$, $a_{nk} \in \mathbb{Q}_p$, $n, k = 0, 1, 2, \ldots$ defined by

$$A \equiv (a_{nk}) = \begin{bmatrix} \frac{1}{r} & \frac{1}{r} & \cdots & \frac{1}{r} & 0 & 0 & 0 & \cdots \\ 0 & \frac{1}{r} & \frac{1}{r} & \cdots & \frac{1}{r} & 0 & 0 & \cdots \\ 0 & 0 & \frac{1}{r} & \frac{1}{r} & \cdots & \frac{1}{r} & 0 & \cdots \\ \cdots & \cdots & \cdots & \cdots & \cdots & \cdots & \cdots & \cdots \end{bmatrix},$$

the entry $\frac{1}{r}$, occurring r times in each row,

$$\left. \begin{aligned} i.e., \; a_{nk} \;\; &= \tfrac{1}{r}, \; k = n, n+1, \ldots, n+r-1 \\ &= 0, \; \text{otherwise} \end{aligned} \right\}, \; n = 0, 1, 2, \ldots.$$

$$(Ax)_{n+1} - (Ax)_n = \frac{(x_{n+1} + x_{n+2} + \cdots + x_{n+r}) - (x_n + x_{n+1} + \cdots + x_{n+r-1})}{r}$$

$$= \frac{x_{n+r} - x_n}{r}$$

$$\to 0, \; n \to \infty,$$

if $x = \{x_k\} \in \Lambda_r$. Thus A sums all sequences in Λ_r. It is clear that A is regular. □

Bibliography

[1] J.D. Hill and H.J. Hamilton. Operation theory and multiple sequence transformations. *Duke Math. J.*, 8:154–162, 1941.

[2] A.F. Monna. Sur le théorème de Banach-Steinhaus. *Indag. Math.*, 25:121–131, 1963.

[3] P.N. Natarajan. The Steinhaus theorem for Toeplitz matrices in non-archimedean fields. *Comment. Math. Prace Mat.*, 20:417–422, 1978.

[4] P.N. Natarajan. A Steinhaus type theorem. *Proc. Amer. Math. Soc.*, 99:559–562, 1987.

[5] P.N. Natarajan. On certain spaces containing the space of Cauchy sequences. *J. Orissa Math. Soc.*, 9:1–9, 1990.

[6] P.N. Natarajan. Criterion for regular matrices in non-archimedean fields. *J. Ramanujan Math. Soc.*, 6:185–195, 1991.

[7] P.N. Natarajan. On sequences of zeros and ones in non-archimedean analysis: A further study. *African Diaspora J. Math.*, 10:49–54, 2010.

[8] P.N. Natarajan. *An introduction to ultrametric summability theory.* Second edition, Springer, 2015.

[9] I. Schur. Über lineare Transformationen in der Theorie der unendlichen Reihen. *J. Reine Angew Math.*, 151:79–111, 1921.

[10] J.J. Sember and A.R. Freedman. On summing of sequences of 0's and 1's. *Rocky Mountain J. Math.*, 11:419–425, 1981.

[11] D. Somasundaram. Some properties of T-matrices over non-archimedean fields. *Publ. Math. Debrecen*, 21:171–177, 1974.

[12] M. Stieglitz and H. Tietz. Matrix transformationen von Folgenräumen eine Ergebnisübersicht. *Math. Z.*, 154:1–16, 1977.

Chapter 3

Matrix Transformations between Some Other Sequence Spaces

3.1 Introduction

In this chapter, K is invariably a complete, non-trivially valued, non-Archimedean field. The main concern of this chapter is the characterization of the matrix class $(\ell_\alpha, \ell_\alpha)$, $\alpha > 0$. When $K = \mathbb{R}$ or $\mathbb{C}$, a complete characterization of the matrix class $(\ell_\alpha, \ell_\beta)$, $\alpha, \beta \geq 2$, does not seem to be available in the literature. The latest result [4] in this direction characterizes only non-negative matrices in $(\ell_\alpha, \ell_\beta)$, when $\alpha \geq \beta > 1$. In case $K = \mathbb{R}$ or $\mathbb{C}$, a known simple sufficient condition [7] for a matrix A to belong to $(\ell_\alpha, \ell_\alpha)$ is

$$A \in (\ell_\infty, \ell_\infty) \cap (\ell_1, \ell_1).$$

Sufficient conditions or necessary conditions for A to belong to the matrix class $(\ell_\alpha, \ell_\beta)$, when $K = \mathbb{R}$ or $\mathbb{C}$, are available in literature (see e.g., [14]). Necessary and sufficient conditions for a matrix $A \in (\ell_1, \ell_1)$ are well known and are due to Mears [8] (for alternate proofs, see Knopp and Lorentz [3], Fridy [1]). The necessary and sufficient conditions mentioned at the outset are obtained in the main Theorem 26 of Section 3.2. Section 3.3 is devoted to a deduction of a result on multiplication of series from Theorem 26. A Mercerian theorem is proved in Section 3.4. In Section 3.5, Steinhaus-type theorems involving the space ℓ_α are studied. Matrices belonging to the class $(\ell(p), \ell_\infty)$ are characterized in Section 3.6.

3.2   Characterization of matrices in $(\ell_\alpha, \ell_\alpha)$, $\alpha > 0$

The main result of the present chapter is the following (see [11]).

Theorem 26. $A = (a_{nk}) \in (\ell_\alpha, \ell_\alpha)$, $\alpha > 0$, *if and only if*

$$\sup_{k \geq 0} \sum_{n=0}^{\infty} |a_{nk}|^\alpha < \infty. \tag{3.1}$$

Proof. Since $|\cdot|$ is a non-Archimedean valuation, we first observe that

$$||a|^\alpha - |b|^\alpha| \leq |a+b|^\alpha \leq |a|^\alpha + |b|^\alpha, \alpha > 0. \tag{3.2}$$

Sufficiency. If $x = \{x_k\} \in \ell_\alpha$, $\sum_{k=0}^{\infty} a_{nk} x_k$ converges, $n = 0, 1, 2, \ldots$, since $x_k \to 0$, $k \to \infty$ and $\sup_{n,k} |a_{nk}| < \infty$ by (3.1). Also,

$$\sum_{n=0}^{\infty} |(Ax)_n|^\alpha \leq \sum_{n=0}^{\infty} \sum_{k=0}^{\infty} |a_{nk}|^\alpha |x_k|^\alpha$$

$$\leq \left(\sum_{k=0}^{\infty} |x_k|^\alpha \right) \left(\sup_{k \geq 0} \sum_{n=0}^{\infty} |a_{nk}|^\alpha \right)$$

$$< \infty,$$

so that $\{(Ax)_n\} \in \ell_\alpha$.

Necessity. Let $A \in (\ell_\alpha, \ell_\alpha)$, $\alpha > 0$. We first note that

$$\sup_{k \geq 0} |a_{nk}|^\alpha = B_n < \infty, \ n = 0, 1, 2, \ldots.$$

Suppose not. Then

$$\sup_{k \geq 0} |a_{mk}|^\alpha = \infty, \text{ for some } m.$$

We can now choose a strictly increasing sequence $\{k(i)\}$ of positive integers such that

$$|a_{m,k(i)}|^{\alpha} > i^2, \; i = 1, 2, \ldots.$$

Define the sequence $x = \{x_k\}$ by

$$\left.\begin{aligned} x_k &= \frac{1}{a_{m,k(i)}}, \text{ if } k = k(i) \\ &= 0, \text{ if } k \neq k(i) \end{aligned}\right\}, \; i = 1, 2, \ldots.$$

Then $\{x_k\} \in \ell_{\alpha}$, for,

$$\sum_{k=0}^{\infty} |x_k|^{\alpha} = \sum_{i=1}^{\infty} |x_{k(i)}|^{\alpha} < \sum_{i=1}^{\infty} \frac{1}{i^2} < \infty,$$

while, $a_{m,k(i)} x_{k(i)} = 1 \nrightarrow 0$, $i \to \infty$, which is a contradiction. Since $(Ax)_n = a_{nk}$ for the sequence $x = \{x_n\}$, $x_n = 0$, $n \neq k$, $x_k = 1$ and $\{(Ax)_n\} \in \ell_{\alpha}$,

$$\mu_k = \sum_{n=0}^{\infty} |a_{nk}|^{\alpha} < \infty, \; k = 0, 1, 2, \ldots.$$

Suppose $\{\mu_k\}$ is unbounded. Choose a positive integer $k(1)$ such that

$$\mu_{k(1)} > 3.$$

Then choose a positive integer $n(1)$ such that

$$\sum_{n=n(1)+1}^{\infty} |a_{n,k(1)}|^{\alpha} < 1,$$

so that

$$\sum_{n=0}^{n(1)} |a_{n,k(1)}|^{\alpha} > 2.$$

More generally, given the positive integers $k(j)$, $n(j)$, $j \leq m - 1$, choose positive integers $k(m)$, $n(m)$ such that $k(m) > k(m-1)$, $n(m) > n(m-1)$,

$$\sum_{n=n(m-2)+1}^{n(m-1)} \sum_{k=k(m)}^{\infty} B_n k^{-2} < 1,$$

$$\mu_{k(m)} > 2 \sum_{n=0}^{n(m-1)} B_n + \rho^{-\alpha} m^2 \left\{ 2 + \sum_{i=1}^{m-1} i^{-2} \mu_{k(i)} \right\},$$

and

$$\sum_{n=n(m)+1}^{\infty} |a_{n,k(m)}|^{\alpha} < \sum_{n=0}^{n(m-1)} B_n,$$

where, since K is non-trivially valued, there exists $\pi \in K$ such that $0 < \rho = |\pi| < 1$. Now,

$$\sum_{n=n(m-1)+1}^{n(m)} |a_{n,k(m)}|^{\alpha}$$

$$= \mu_{k(m)} - \sum_{n=0}^{n(m-1)} |a_{n,k(m)}|^{\alpha}$$

$$- \sum_{n=n(m)+1}^{\infty} |a_{n,k(m)}|^{\alpha}$$

$$> 2 \sum_{n=0}^{n(m-1)} B_n + \rho^{-\alpha} m^2 \left\{ 2 + \sum_{i=1}^{m-1} i^{-2} \mu_{k(i)} \right\}$$

$$- \sum_{n=0}^{n(m-1)} B_n - \sum_{n=0}^{n(m-1)} B_n$$

$$= \rho^{-\alpha} m^2 \left\{ 2 + \sum_{i=1}^{m-1} i^{-2} \mu_{k(i)} \right\}.$$

For every $i = 1, 2, \ldots$, there exists a non-negative integer $\lambda(i)$ such that

$$\rho^{\lambda(i)+1} \leq i^{-\frac{2}{\alpha}} < \rho^{\lambda(i)}.$$

Define the sequence $x = \{x_k\}$ as follows:

$$\left. \begin{array}{l} x_k = \pi^{\lambda(i)+1}, \text{ if } k = k(i) \\[2mm] = 0, \text{ if } k \neq k(i) \end{array} \right\}, \ i = 1, 2, \ldots.$$

Then $\{x_k\} \in \ell_\alpha$, for,

$$\sum_{k=0}^{\infty} |x_k|^{\alpha} = \sum_{i=1}^{\infty} |x_{k(i)}|^{\alpha} \leq \sum_{i=1}^{\infty} \frac{1}{i^2} < \infty.$$

However, using (3.2), we have,

$$\sum_{n=n(m-1)+1}^{n(m)} |(Ax)_n|^{\alpha} \geq \Sigma_1 - \Sigma_2 - \Sigma_3,$$

where

$$\Sigma_1 = \sum_{n=n(m-1)+1}^{n(m)} |a_{n,k(m)}|^\alpha |x_{k(m)}|^\alpha,$$

$$\Sigma_2 = \sum_{n=n(m-1)+1}^{n(m)} \sum_{i=1}^{m-1} |a_{n,k(i)}|^\alpha |x_{k(i)}|^\alpha,$$

$$\Sigma_3 = \sum_{n=n(m-1)+1}^{n(m)} \sum_{i=m+1}^{\infty} |a_{n,k(i)}|^\alpha |x_{k(i)}|^\alpha.$$

Now,

$$\Sigma_1 = \sum_{n=n(m-1)+1}^{n(m)} |a_{n,k(m)}|^\alpha \rho^{(\lambda(m)+1)\alpha}$$

$$\geq \rho^\alpha \sum_{n=n(m-1)+1}^{n(m)} |a_{n,k(m)}|^\alpha m^{-2}$$

$$> 2 + \sum_{i=1}^{m-1} i^{-2} \mu_{k(i)}; \tag{3.3}$$

$$\Sigma_2 = \sum_{n=n(m-1)+1}^{n(m)} \sum_{i=1}^{m-1} |a_{n,k(i)}|^\alpha \rho^{(\lambda(i)+1)\alpha}$$

$$\leq \sum_{i=1}^{m-1} i^{-2} \sum_{n=n(m-1)+1}^{n(m)} |a_{n,k(i)}|^\alpha$$

$$\leq \sum_{i=1}^{m-1} i^{-2} \mu_{k(i)}; \tag{3.4}$$

$$\Sigma_3 = \sum_{n=n(m-1)+1}^{n(m)} \sum_{i=m+1}^{\infty} |a_{n,k(i)}|^\alpha \rho^{(\lambda(i)+1)\alpha}$$

$$\leq \sum_{n=n(m-1)+1}^{n(m)} \sum_{i=k(m+1)}^{\infty} B_n i^{-2}$$

$$< 1. \tag{3.5}$$

Thus, from (3.3)-(3.5), we have,

$$\sum_{n=n(m-1)+1}^{n(m)} |(Ax)_n|^\alpha > 1, \ m = 2, 3, \ldots.$$

This shows that $\{(Ax)_n\} \notin \ell_\alpha$, while $x = \{x_k\} \in \ell_\alpha$, a contradiction. Thus condition (3.1) is also necessary. This completes the proof of the theorem. □

3.3   Multiplication of series

In this section, we consider the Cauchy product of two series $\sum\limits_{k=0}^{\infty} a_k, \sum\limits_{k=0}^{\infty} b_k$, viz., $\sum\limits_{k=0}^{\infty} c_k$, where

$$c_k = a_0 b_k + a_1 b_{k-1} + \cdots + a_k b_0, \ k = 0, 1, 2, \ldots. \tag{3.6}$$

In the context of the sequence space ℓ_α, we have the following result.

Theorem 27. *If $a = \{a_k\}$, $b = \{b_k\} \in \ell_\alpha$, $\alpha > 0$, so is their Cauchy product $c = \{c_k\}$.*

Proof. Consider the matrix

$$A \equiv \begin{bmatrix} a_0 & 0 & 0 & 0 & \ldots \\ a_1 & a_0 & 0 & 0 & \ldots \\ a_2 & a_1 & a_0 & 0 & \ldots \\ \ldots & \ldots & \ldots & \ldots & \ldots \end{bmatrix}. \tag{3.7}$$

Noting that the A-transform of $\{b_k\}$ is $\{c_k\}$, since $a = \{a_k\} \in \ell_\alpha$, $A \in (\ell_\alpha, \ell_\alpha)$ so that $c = \{c_k\} \in \ell_\alpha$, since $b = \{b_k\} \in \ell_\alpha$. □

Remark 7. *(i) It is easy to establish Theorem 27, when $\alpha = 1$, by virtually following the steps in the well-known Cauchy theorem on multiplication of series. Theorem 27 could be proved in the same way using (3.2).*

(ii) Theorem 27 could be stated formally in the following form too: If a sequence $\{a_k\}$ is given, then $\{c_k\} \in \ell_\alpha$ for every sequence $\{b_k\} \in \ell_\alpha$ if and only if $\{a_k\} \in \ell_\alpha$, where c_k is defined by (3.6). To establish the necessity part of this

assertion, we need only observe that the condition $\sum_{k=0}^{\infty} |a_k|^{\alpha} < \infty$ *is necessary for the matrix A in (3.7) to belong to $(\ell_\alpha, \ell_\alpha)$.*

(iii) Theorem 27 is not true when $K = \mathbb{R}$ or $\mathbb{C}$ as illustrated by the following example: Let

$$a_k = b_k = \frac{1}{(k+1)^{\frac{1}{2}(1+\frac{1}{\alpha})}}, \ \alpha > 1.$$

Then

$$\{a_k\}, \{b_k\} \in \ell_\alpha,$$

for,

$$\sum_{k=0}^{\infty} |a_k|^{\alpha} = \sum_{k=0}^{\infty} |b_k|^{\alpha}$$

$$= \sum_{k=0}^{\infty} \frac{1}{(k+1)^{\frac{\alpha}{2}(1+\frac{1}{\alpha})}}$$

$$< \sum_{k=0}^{\infty} \frac{1}{k^{\frac{\alpha+1}{2}}}$$

$$< \infty,$$

since $\alpha > 1$. If $0 \leq 1 \leq k$, $a_k b_{k-1} \geq \frac{1}{(k+1)^{1+\frac{1}{\alpha}}}$, so that

$$c_k \geq \frac{k+1}{(k+1)^{1+\frac{1}{\alpha}}} = \frac{1}{(k+1)^{\frac{1}{\alpha}}}.$$

Hence

$$\sum_{k=0}^{\infty} |c_k|^{\alpha} \geq \sum_{k=0}^{\infty} \frac{1}{k+1} = \infty,$$

and so $\{c_k\} \notin \ell_\alpha$.

It is not Theorem 27, but the following Theorem 28 which reveals the real situation in the non-Archimedean setup (see [10]).

Theorem 28. *If $\sum_{k=0}^{\infty} a_k$, $\sum_{k=0}^{\infty} b_k$ are two infinite series, then $\sum_{k=0}^{\infty} c_k$, where c_k is defined by (3.6), converges for every convergent series $\sum_{k=0}^{\infty} a_k$ if and only if $\sum_{k=0}^{\infty} b_k$ converges.*

Proof. Let $\displaystyle\sum_{k=0}^{\infty} b_k$ be given such that $\displaystyle\sum_{k=0}^{\infty} c_k$ converges for every convergent series $\displaystyle\sum_{k=0}^{\infty} a_k$. For the series $\displaystyle\sum_{k=0}^{\infty} a_k$, $a_0 = 1$, $a_k = 0$, $k = 1, 2, \ldots$, $c_k = b_k$ and so $\displaystyle\sum_{k=0}^{\infty} b_k$ converges. Conversely, let $\displaystyle\sum_{k=0}^{\infty} a_k$, $\displaystyle\sum_{k=0}^{\infty} b_k$ converge. There exists $M > 0$ such that

$$|a_k|, |b_k| < M, \ k = 0, 1, 2, \ldots.$$

Given $\epsilon > 0$, there exists a positive integer N_1 such that

$$|a_k| < \frac{\epsilon}{M}, \quad \text{for all } k > N_1.$$

Since $\lim_{k \to \infty} b_{k-r} = 0$, $r = 0, 1, 2, \ldots$, we can choose a positive integer $N_2 > N_1$ such that

$$\sup_{0 \le r \le N_1} |b_{k-r}| < \frac{\epsilon}{M}, \text{ for } k > N_2.$$

Then, for $k > N_2$,

$$
\begin{aligned}
|c_k| &= \left| \sum_{r=0}^{k} b_{k-r} a_r \right| \\
&= \left| \sum_{r=0}^{N_1} b_{k-r} a_r + \sum_{r=N_1+1}^{k} b_{k-r} a_r \right| \\
&\le \max \left\{ \sup_{0 \le r \le N_1} |b_{k-r}||a_r|, \ \sup_{N_1+1 \le r \le k} |b_{k-r}||a_r| \right\} \\
&< \max \left\{ \frac{\epsilon}{M} M, M \frac{\epsilon}{M} \right\} \\
&= \epsilon.
\end{aligned}
$$

Thus $\displaystyle\sum_{k=0}^{\infty} c_k$ converges. It is easy to check that

$$\sum_{k=0}^{\infty} c_k = \left(\sum_{k=0}^{\infty} a_k \right) \left(\sum_{k=0}^{\infty} b_k \right).$$

This completes the proof of the theorem.   □

3.4 A Mercerian theorem

We now study the structure of the matrix class $(\ell_\alpha, \ell_\alpha)$, $\alpha \geq 1$, with a view to obtain a Mercerian theorem analogous to the one obtained earlier by Rangachari and Srinivasan [13]. $(\ell_\alpha, \ell_\alpha)$ is a Banach algebra under the norm

$$\|A\| = \sup_{k \geq 0} \left(\sum_{n=0}^{\infty} |a_{nk}|^\alpha \right)^{\frac{1}{\alpha}}, \; A \equiv (a_{nk}) \in (\ell_\alpha, \ell_\alpha), \tag{3.8}$$

with the usual matrix addition, multiplication and scalar multiplication. First we note that if $A = (a_{nk})$, $B = (b_{nk}) \in (\ell_\alpha, \ell_\alpha)$, $(AB)_{nk}$ is defined, for, $(AB)_{nk} = \sum_{i=0}^{\infty} a_{ni} b_{ik}$ converges, since $b_{ik} \to 0$, $i \to \infty$ and $\sup_{n,i} |a_{ni}| < \infty$. We next show that $(\ell_\alpha, \ell_\alpha)$ is closed with respect to multiplication. For, using (3.2),

$$\sum_{n=0}^{\infty} |(AB)_{nk}|^\alpha = \sum_{n=0}^{\infty} \left| \sum_{i=0}^{\infty} a_{ni} b_{ik} \right|^\alpha$$

$$\leq \sum_{n=0}^{\infty} \sum_{i=0}^{\infty} |a_{ni}|^\alpha |b_{ik}|^\alpha$$

$$= \left(\sum_{i=0}^{\infty} |b_{ik}|^\alpha \right) \left(\sum_{n=0}^{\infty} |a_{ni}|^\alpha \right)$$

$$\leq \|A\|^\alpha \|B\|^\alpha, \; k = 0, 1, 2, \ldots.$$

Thus $AB \in (\ell_\alpha, \ell_\alpha)$ and $\|AB\| \leq \|A\|\|B\|$. The associative law follows, for, if $A = (a_{nk})$, $B = (b_{nk})$, $c = (c_{nk}) \in (\ell_\alpha, \ell_\alpha)$,

$$((AB)C)_{nk} = \sum_{i=0}^{\infty} (AB)_{ni} c_{ik}$$

$$= \sum_{i=0}^{\infty} c_{ik} \left(\sum_{j=0}^{\infty} a_{nj} b_{ji} \right)$$

$$= \sum_{j=0}^{\infty} a_{nj} \left(\sum_{i=0}^{\infty} b_{ji} c_{ik} \right)$$

$$= \sum_{j=0}^{\infty} a_{nj} (BC)_{jk}$$

$$= (A(BC))_{nk}, \ n, k = 0, 1, 2, \ldots.$$

It remains to prove that $(\ell_\alpha, \ell_\alpha)$ is complete under the norm defined by (3.8). To prove this, let $\{A^{(n)}\}_{n=0}^{\infty}$ be a Cauchy sequence in $(\ell_\alpha, \ell_\alpha)$, where

$$A^{(n)} = (a_{ij}^{(n)}), \ i, j = 0, 1, 2, \ldots.$$

Since $\{A^{(n)}\}_{n=0}^{\infty}$ is Cauchy, for any $\epsilon > 0$, there exists a positive integer n_0 such that for $m, n \geq n_0$,

$$\|A^{(m)} - A^{(n)}\| < \epsilon,$$

$$i.e., \ \sup_{j \geq 0} \sum_{i=0}^{\infty} |a_{ij}^{(m)} - a_{ij}^{(n)}|^\alpha < \epsilon^\alpha.$$

Thus, for all $i, j = 0, 1, 2, \ldots,$

$$|a_{ij}^{(m)} - a_{ij}^{(m)}| < \epsilon, \ m, n \geq n_0.$$

Since K is complete,

$$a_{ij}^{(n)} \to a_{ij}, \ n \to \infty, i, j = 0, 1, 2, \ldots.$$

Consider the matrix $A = (a_{ij})$, $i, j = 0, 1, 2, \ldots.$ Now, for all $j = 0, 1, 2, \ldots,$

$$\sum_{i=0}^{k} |a_{ij}^{(n)}|^\alpha \leq \|A^{(n)}\|^\alpha \leq M^\alpha, \ n, k = 0, 1, 2, \ldots,$$

where $M = \sup\limits_{n \geq 0} \|A^{(n)}\|$. Allowing $n \to \infty$, we have,

$$\sum_{i=0}^{k} |a_{ij}|^\alpha \leq M^\alpha, \; j, k = 0, 1, 2, \ldots.$$

Allowing $k \to \infty$,

$$\sum_{i=0}^{\infty} |a_{ij}|^\alpha \leq M^\alpha, \; j = 0, 1, 2, \ldots,$$

which shows that $A \in (\ell_\alpha, \ell_\alpha)$. Again, for all $j, k = 0, 1, 2, \ldots,$

$$\sum_{i=0}^{k} |a_{ij}^{(m)} - a_{ij}^{(n)}|^\alpha < \epsilon^\alpha, \; m, n \geq n_0.$$

For $n \geq n_0$, allowing $m \to \infty$, for all $j, k = 0, 1, 2, \ldots,$

$$\sum_{i=0}^{k} |a_{ij} - a_{ij}^{(n)}|^\alpha \leq \epsilon^\alpha.$$

Now, allowing $k \to \infty$, we have, for all $j = 0, 1, 2, \ldots,$

$$\sum_{i=0}^{\infty} |a_{ij} - a_{ij}^{(n)}|^\alpha \leq \epsilon^\alpha, \; n \geq n_0,$$

$$i.e., \; \sup_{j \geq 0} \left(\sum_{i=0}^{\infty} |a_{ij} - a_{ij}^{(n)}|^\alpha \right)^{\frac{1}{\alpha}} \leq \epsilon, \; n \geq n_0,$$

$$i.e., \; \|A^{(n)} - A\| \leq \epsilon, \; n \geq n_0,$$

which shows that $A^{(n)} \to A, \; n \to \infty$. The proof of the theorem is now complete.

The Mercerian theorem mentioned is the following (see [11]).

Theorem 29. *When $K = \mathbb{Q}_p$, the p-adic field for a prime p, if*

$$y_n = x_n + \lambda p^n (x_0 + x_1 + \cdots + x_n)$$

and $\{y_n\} \in \ell_\alpha$, then $\{x_n\} \in \ell_\alpha$ if $|\lambda|_p < (1 - \rho^\alpha)^{\frac{1}{\alpha}}$, where $\rho = |p|_p < 1$.

Proof. Since $(\ell_\alpha, \ell_\alpha)$ is a Banach algebra, if $\lambda \in \mathbb{Q}_p$ is such that $|\lambda|_p < \frac{1}{\|A\|}$,

$A \in (\ell_\alpha, \ell_\alpha)$, then $I - \lambda A$, where I is the identity matrix, has an inverse in $(\ell_\alpha, \ell_\alpha)$. The matrix of the transformation is $I + \lambda A$, where

$$
A \equiv \begin{bmatrix}
1 & 0 & 0 & 0 & \cdots \\
p & p & 0 & 0 & \cdots \\
p^2 & p^2 & p^2 & 0 & \cdots \\
\cdots & \cdots & \cdots & \cdots & \cdots
\end{bmatrix}.
$$

We note that $A \in (\ell_\alpha, \ell_\alpha)$, with $\|A\| = \dfrac{1}{(1-\rho^\alpha)^{\frac{1}{\alpha}}}$. Then $I + \lambda A$ has an inverse in $(\ell_\alpha, \ell_\alpha)$ if $|\lambda|_p < (1 - \rho^\alpha)^{\frac{1}{\alpha}}$. Since $y = (I + \lambda A)x$, where $y = \{y_k\}$, $x = \{x_k\}$ and lower triangular matrices are associative, it follows that

$$
(I + \lambda A)^{-1} y = x.
$$

Since $y \in \ell_\alpha$ and $(I + \lambda A)^{-1} \in (\ell_\alpha, \ell_\alpha)$, it follows that $x \in \ell_\alpha$, completing the proof. □

3.5　Another Steinhaus-type theorem

We establish in Theorem 31, to be proved in the sequel, a Steinhaus-type result, using the characterization of $(\ell_\alpha, \ell_\alpha)$ in Theorem 26.

We write

$$
A \in (\ell_\alpha, \ell_\alpha; P)
$$

if $A \in (\ell_\alpha, \ell_\alpha)$ and $\displaystyle\sum_{n=0}^{\infty} (Ax)_n = \sum_{k=0}^{\infty} x_k$, $x = \{x_k\} \in \ell_\alpha$;

$$
A \in (\ell_\alpha, \ell_\alpha; P)'
$$

if $A \in (\ell_\alpha, \ell_\alpha; P)$ and $a_{nk} \to 0$, $k \to \infty$, $n = 0, 1, 2, \ldots$. It is easy to prove that $A \in (\ell_\alpha, \ell_\alpha; P)$ if and only if

$$
A \in (\ell_\alpha, \ell_\alpha) \ \text{ and } \ \sum_{n=0}^{\infty} a_{nk} = 1, \ k = 0, 1, 2, \ldots.
$$

Theorem 31 is an immediate consequence of the following result (see [11]).

Theorem 30. *If $A \in (\ell_\alpha, \ell_\alpha)$ such that $a_{nk} \to 0$, $k \to \infty$, $n = 0, 1, 2, \ldots$ and $\overline{\lim\limits_{k \to \infty}} \sum\limits_{n=0}^{\infty} |a_{nk}|^\alpha > 0$, then there exists a sequence $x = \{x_k\} \in \ell_\beta$, $\beta > \alpha$, $Ax = \{(Ax)_n\} \notin \ell_\alpha$.*

Proof. By hypothesis, for some $\epsilon > 0$, there exists a sub-sequence $\{k(i)\}$ of positive integers such that

$$\sum_{n=0}^{\infty} |a_{n,k(i)}|^\alpha \geq 2\epsilon, i = 1, 2, \ldots.$$

In particular,

$$\sum_{n=0}^{\infty} |a_{n,k(1)}|^\alpha \geq 2\epsilon.$$

Choose a positive integer $n(1)$ such that

$$\sum_{n=n(1)+1}^{\infty} |a_{n,k(1)}|^\alpha < \min(2^{-1}, \frac{\epsilon}{2}),$$

so that

$$\sum_{n=0}^{n(1)} |a_{n,k(1)}|^\alpha > \epsilon.$$

More generally, having chosen the positive integers $k(j)$, $n(j)$, $j \leq m - 1$, choose a positive integer $k(m)$ such that $k(m) > k(m-1)$,

$$\sum_{n=0}^{\infty} |a_{n,k(m)}|^\alpha \geq 2\epsilon,$$

$$\sum_{n=0}^{n(m-1)} |a_{n,k(m)}|^\alpha < \min(2^{-m}, \frac{\epsilon}{2}),$$

and then choose a positive integer $n(m)$ such that $n(m) > n(m-1)$,

$$\sum_{n=n(m)+1}^{\infty} |a_{n,k(m)}|^\alpha < \min(2^{-m}, \frac{\epsilon}{2}),$$

so that

$$\sum_{n=n(m-1)+1}^{n(m)} |a_{n,k(m)}|^\alpha > 2\epsilon - \frac{\epsilon}{2} - \frac{\epsilon}{2}$$

$$= \epsilon.$$

Since K is non-trivially valued, there exists $\pi \in K$ such that $0 < \rho = |\pi| < 1$. For each $i = 1, 2, \ldots$, choose a non-negative integer $\lambda(i)$ such that

$$\rho^{\lambda(i)+1} \leq \frac{1}{i^{\frac{1}{\alpha}}} < \rho^{\lambda(i)}.$$

Define the sequence $x = \{x_k\}$ by

$$\left. \begin{aligned} x_k \quad &= \pi^{\lambda(i)}, \text{ if } k = k(i) \\ &= 0, \text{ if } k \neq k(i) \end{aligned} \right\} , \quad i = 1, 2, \ldots.$$

Then, $\{x_k\} \in \ell_\beta - \ell_\alpha$, for,

$$\sum_{k=0}^{\infty} |x_k|^\beta = \sum_{i=1}^{\infty} |x_{k(i)}|^\beta$$

$$= \sum_{i=1}^{\infty} \rho^{\beta\lambda(i)}$$

$$\leq \frac{1}{\rho^\beta} \sum_{i=1}^{\infty} \frac{1}{i^{\frac{\beta}{\alpha}}}$$

$$< \infty,$$

since $\beta > \alpha$, while,

$$\sum_{k=0}^{\infty} |x_k|^\alpha = \sum_{i=1}^{\infty} |x_{k(i)}|^\alpha$$

$$= \sum_{i=1}^{\infty} \rho^{\alpha\lambda(i)}$$

$$> \sum_{i=1}^{\infty} \frac{1}{i}$$

$$= \infty.$$

Defining $n(0) = 0$,

$$\sum_{n=0}^{n(N)} |(Ax)_n|^\alpha \geq \sum_{m=1}^{N} \sum_{n=n(m-1)+1}^{n(m)} \left| \sum_{i=1}^{\infty} a_{n,k(i)} x_{k(i)} \right|^\alpha$$

$$= \sum_{m=1}^{N} \sum_{n=n(m-1)+1}^{n(m)} \left| \sum_{i=1}^{\infty} a_{n,k(i)} \pi^{\lambda(i)} \right|^\alpha$$

$$\geq \sum_{m=1}^{N} \sum_{n=n(m-1)+1}^{n(m)} \left\{ |a_{n,k(m)}|^\alpha \rho^{\alpha\lambda(m)} \right.$$

$$\left. - \sum_{i \neq m} |a_{n,k(i)}|^\alpha \rho^{\alpha\lambda(i)} \right\},$$

using (3.2)

$$\geq \sum_{m=1}^{N} \sum_{n=n(m-1)+1}^{n(m)} \left\{ |a_{n,k(m)}|^\alpha m^{-1} \right.$$

$$\left. - \frac{1}{\rho^\alpha} \sum_{i \neq m} |a_{n,k(i)}|^\alpha \right\},$$

since $\rho^{\alpha(\lambda(i)+1)} \leq \dfrac{1}{i} \leq 1$ so that $\rho^{\alpha\lambda(i)} \leq \dfrac{1}{\rho^\alpha}$

$$> \sum_{m=1}^{N} \epsilon m^{-1} - \frac{1}{\rho^\alpha} \sum_{m=1}^{N} \sum_{n=n(m-1)+1}^{n(m)} \sum_{i \neq m} |a_{n,k(i)}|^\alpha. \qquad (3.9)$$

We note that

$$\sum_{m=1}^{\infty} \sum_{n=n(m-1)+1}^{n(m)} \sum_{i<m} |a_{n,k(i)}|^\alpha$$

$$= \sum_{m=1}^{\infty} \sum_{n=n(m)+1}^{\infty} |a_{n,k(m)}|^\alpha$$

$$< \sum_{m=1}^{\infty} 2^{-m}. \qquad (3.10)$$

Similarly, it can be proved that

$$\sum_{m=1}^{\infty} \sum_{n=n(m-1)+1}^{n(m)} \sum_{i>m} |a_{n,k(i)}|^\alpha < \sum_{m=1}^{\infty} 2^{-(m+1)}. \qquad (3.11)$$

Thus, it follows from (3.9)-(3.11) that

$$\sum_{n=0}^{n(N)} |(Ax)_n|^\alpha > \epsilon \sum_{m=1}^{N} \frac{1}{m} - \frac{3}{2}.$$

Since $\sum_{m=1}^{\infty} \frac{1}{m} = \infty$, it follows that $\{(Ax)_n\} \notin \ell_\alpha$, completing the proof of the theorem. $\qquad\square$

It is now easy to prove the following:

Theorem 31. $(\ell_\alpha, \ell_\alpha; P)' \cap (\ell_\beta, \ell_\alpha) = \phi$, $\beta > \alpha$.

Proof. Suppose $A \equiv (a_{nk}) \in (\ell_\alpha, \ell_\alpha; P)' \cap (\ell_\beta, \ell_\alpha)$, $\beta > \alpha$. Then

$$\sum_{n=0}^{\infty} |a_{nk}|^\alpha \geq \left| \sum_{n=0}^{\infty} a_{nk} \right|^\alpha = 1, \ k = 0, 1, 2, \ldots,$$

so that

$$\varlimsup_{k \to \infty} \sum_{n=0}^{\infty} |a_{nk}|^\alpha \geq 1.$$

In view of Theorem 30, there exists $x = \{x_k\} \in \ell_\beta$ such that $\{(Ax)_n\} \notin \ell_\alpha$, completing the proof. $\qquad\square$

Remark 8. *(i) When $K = \mathbb{R}$ or $\mathbb{C}$, it was proved by Fridy [2] that*

$$(\ell_1, \ell_1; P) \cap (\ell_\alpha, \ell_1) = \phi, \ \alpha > 1.$$

This result, as such, fails to hold when K is a complete, non-trivially valued, non-Archimedean field as the following example shows: Let $K = \mathbb{Q}_3$ and A be the infinite matrix

$$A \equiv (a_{nk}) = \begin{bmatrix} \frac{1}{4} & \frac{1}{4} & \frac{1}{4} & \cdots \\ \frac{1}{4}\left(\frac{3}{4}\right) & \frac{1}{4}\left(\frac{3}{4}\right) & \frac{1}{4}\left(\frac{3}{4}\right) & \cdots \\ \frac{1}{4}\left(\frac{3}{4}\right)^2 & \frac{1}{4}\left(\frac{3}{4}\right)^2 & \frac{1}{4}\left(\frac{3}{4}\right)^2 & \cdots \\ \cdots & \cdots & \cdots & \cdots \end{bmatrix},$$

i.e., $a_{nk} = \frac{1}{4}\left(\frac{3}{4}\right)^n$, $k = 0, 1, 2, \ldots, n = 0, 1, 2, \ldots$.

Now,

$$\sup_{k \geq 0} \sum_{n=0}^{\infty} |a_{nk}|_3 = \frac{1}{1-\rho} < \infty, \quad where \ \rho = |3|_3$$

and

$$\sum_{n=0}^{\infty} a_{nk} = 1, \ k = 0, 1, 2, \ldots,$$

so that $A \in (\ell_1, \ell_1; P)$; but, for $\alpha > 1$, if $x = \{x_k\} \in \ell_\alpha \subset c_0$,

$$\sum_{n=0}^{\infty} |(Ax)_n|_3 = \sum_{n=0}^{\infty} \left| \sum_{k=0}^{\infty} a_{nk} x_k \right|_3$$

$$= \sum_{n=0}^{\infty} \left| \frac{1}{4} \left(\frac{3}{4} \right)^n \right|_3 \left| \sum_{k=0}^{\infty} x_k \right|_3$$

$$= \left| \sum_{k=0}^{\infty} x_k \right|_3 \frac{1}{1-\rho}$$

$$< \infty,$$

showing that $A \in (\ell_\alpha, \ell_1)$ too:

(ii) It is clear that $(\ell_\alpha, \ell_\alpha; P) \cap (\ell_\infty, \ell_\alpha) = \phi$, since, if $A \in (\ell_\alpha, \ell_\alpha; P) \cap (\ell_\infty, \ell_\alpha)$, then $A \in (\ell_\alpha, \ell_\alpha; P)' \cap (\ell_\beta, \ell_\alpha)$, $\beta > \alpha$, a contradiction.

(iii) By virtually following the proof of Theorem 30, we can show that given any matrix $A \in (\ell_\alpha, \ell_\alpha; P)$, there exists a sequence of 0's and 1's, whose A-transform is not in ℓ_α. This is analogous to Schur's version of the Steinhaus theorem for regular matrices.

3.6   Characterization of matrices in $(\ell(p), \ell_\infty)$

In this section, we characterize the matrix class $(\ell(p), \ell_\infty)$, where $\ell(p)$ is defined (Definition 20) with reference to a sequence $p = \{p_k\}$ of positive real numbers such that

$$0 < p_k \leq \sup_{k \geq 0} p_k < \infty,$$

in the context of a complete, non-trivially valued, non-Archimedean field K. The paranorm on $\ell(p)$ is given by (1.8). The following theorem is the analogue of a result of Lascarides and Maddox [5] corresponding to $K = \mathbb{R}$ or $\mathbb{C}$.

Theorem 32 (see [12]). *If $A \equiv (a_{nk})$, $a_{nk} \in K$, $n, k = 0, 1, 2, \dots$ is an infinite matrix, then $A \in (\ell(p), \ell_\infty)$ if and only if*

$$M \equiv \sup_{n,k} |a_{nk}|^{p_k} < \infty. \tag{3.12}$$

Proof. Let $\ell^\times(p)$ be the generalized Köthe-Toeplitz dual of $\ell(p)$ as in Definition 21. If $\{x_k\} \in \ell_\infty(p)$ (see Definition 20), $\{y_k\} \in \ell(p)$, then

$$|x_k y_k|^{p_k} \to 0, \ k \to \infty.$$

Since K is non-trivially valued,

$$|x_k y_k| \to 0, \ k \to \infty$$

and so $\displaystyle\sum_{k=0}^{\infty} x_k y_k$ converges. Thus, $\ell_\infty(p) \subset \ell^\times(p)$. If, now, $\{x_k\} \in \ell^\times(p)$, then

$$x_k y_k \to 0, \ k \to \infty, \ \text{whenever} \ \{y_k\} \in \ell(p).$$

However, if $\{x_k\} \notin \ell_\infty(p)$, we can find a strictly increasing sequence $\{k(i)\}$ of positive integers such that

$$|x_{k(i)}|^{p_{k(i)}} \geq i^2, \ i = 1, 2, \dots .$$

The sequence $\{y_k\}$ defined by

$$\left. \begin{array}{l} y_k \ = x_k^{-1}, \ \text{if} \ k = k(i) \\[2mm] \quad = 0, \ \text{if} \ k \neq k(i) \end{array} \right\} , \ i = 1, 2, \dots,$$

is in $\ell(p)$, for,

$$\sum_{k=0}^{\infty} |y_k|^{p_k} = \sum_{i=1}^{\infty} |y_{k(i)}|^{p_{k(i)}}$$

$$= \sum_{i=1}^{\infty} |x_{k(i)}|^{-p_{k(i)}}$$

$$\leq \sum_{i=1}^{\infty} \frac{1}{i^2}$$

$$< \infty.$$

However, $x_k y_k \nrightarrow 0$, $k \to \infty$. This is a contradiction. Thus $\{x_k\} \in \ell_{\infty}(p)$. In other words, $\ell^{\times}(p) = \ell_{\infty}(p)$ (cf. Maddox [6]).

Sufficiency. Suppose (3.12) holds. $\{a_{nk}\}_{k=0}^{\infty} \in \ell_{\infty}(p) = \ell^{\times}(p)$, $n = 0, 1, 2, \ldots$. This ensures the existence of the A-transform $\{(Ax)_n\}$ of any $x = \{x_k\} \in \ell(p)$. If $x = \{x_k\} \in \ell(p)$, $|x_k|^{p_k} \to 0$, $k \to \infty$, so that there exists a positive integer N such that for $k > N$,

$$|x_k|^{p_k} < \frac{1}{M},$$

$$i.e., M^{\frac{1}{p_k}} |x_k| < 1,$$

where M is as in (3.12). Hence,

$$|(Ax)_n| \leq \sup_{k \geq 0} |a_{nk}||x_k|$$

$$\leq \sup_{k \geq 0} M^{\frac{1}{p_k}} |x_k|$$

$$\leq \max \left(\max_{0 \leq k \leq N} M^{\frac{1}{p_k}} |x_k|, 1 \right),$$

which means that $\{(Ax)_n\} \in \ell_{\infty}$.

Necessity. Let $A \in (\ell(p), \ell_{\infty})$. This implies that $\{a_{nk}\}_{k=0}^{\infty} \in \ell^{\times}(p) = \ell_{\infty}(p)$, $n = 0, 1, 2, \ldots$. The transformation

$$A_n : \ell(p) \to K$$

defined by $A_n(x) = (Ax)_n$ is a continuous linear functional on $\ell(p)$ for each $n = 0, 1, 2, \ldots$. To see this, let

$$L_n = \sup_{k \geq 0} |a_{nk}|^{p_k}, \; n = 0, 1, 2, \ldots,$$

noting that $L_n < \infty$, since $\{a_{nk}\}_{k=0}^{\infty} \in \ell_{\infty}(p)$, $n = 0, 1, 2, \ldots$. We can suppose, as we may, $L_n > 1$, $n = 0, 1, 2, \ldots$. Hence

$$|a_{nk}| L_n^{-\frac{1}{p_k}} \leq 1, \; k = 0, 1, 2, \ldots$$

and so

$$|A_n(x)| \leq \sup_{k \geq 0} |a_{nk}||x_k|$$
$$= \sup_{k \geq 0} \{|a_{nk}| L_n^{-\frac{1}{p_k}} L_n^{\frac{1}{p_k}} |x_k|\}.$$

If we choose $g(x) \leq \frac{1}{L_n}$,

$$|x_k|^{p_k} \leq \sum_{k=0}^{\infty} |x_k|^{p_k}$$
$$= \{g(x)\}^H$$
$$\leq \frac{1}{L_n^H},$$
i.e., $L_n^{\frac{H}{p_k}} |x_k| \leq 1, \; k = 0, 1, 2, \ldots$.

Consequently,

$$|A_n(x)| \leq \sup_{k \geq 0} L_n^{\frac{H}{p_k}} |x_k|$$
$$\leq \sup_{k \geq 0} L_n |x_k|^{\frac{p_k}{H}}, \quad \text{since } L_n^{\frac{H}{p_k}} |x_k| \leq 1$$
$$\text{and } \frac{p_k}{H} \leq 1$$
$$\leq L_n g(x),$$

which proves the continuity of A_n. Hence by the Uniform Boundedness Principle [9], there exists $\delta > 0$ ($\delta < 1$, as we may assume) and $G \geq 1$ such that

$$|A_n(x)| \leq G, \; n = 0, 1, 2, \ldots$$

whenever $x \in S_\delta(0) = \{x = \{x_k\} : g(x) \le \delta\}$.

Since K is non-trivially valued, there exists $\pi \in K$ such that $0 < \rho = |\pi| < 1$. For each $k = 0, 1, 2, \ldots$, choose a non-negative integer $\lambda(k)$ such that

$$\rho^{\lambda(k)+1} \le \delta^{\frac{H}{p_k}} < \rho^{\lambda(k)}.$$

If we define $x = \{x_n\}$, where

$$x_n \quad = 0, \text{ if } n \ne k;$$

$$= \pi^{\lambda(k)+1}, \text{ if } n = k,$$

we have,

$$g(x) = \rho^{\{\lambda(k)+1\}\frac{p_k}{H}} \le \delta,$$

so that $x \in S_\delta(0)$ implying $|A_n(x)| \le G$. This means

$$|a_{nk}| \le G\rho^{-\{\lambda(k)+1\}},$$

this being true for all $n, k = 0, 1, 2, \ldots$. Thus

$$|a_{nk}|^{p_k} \le G^{p_k} \rho^{-p_k \lambda(k)} \rho^{-p_k}$$

$$\le G^H \delta^{-H} \rho^{-H},$$

proving that

$$\sup_{n,k} |a_{nk}|^{p_k} < \infty,$$

completing the proof of the theorem.   □

Remark 9. *(i) It is easily seen that $\ell_\infty(p) = \ell_\infty$ if and only if $\inf_{k \ge 0} p_k > 0$ (cf. Section 1.6). In this case, the condition*

$$\sup_{n,k} |a_{nk}|^{p_k} < \infty$$

is equivalent to the condition

$$\sup_{n,k} |a_{nk}| < \infty.$$

Because $c_0^\times(p) = \ell_\infty(p)$ (see Section 1.6), one implication of Theorem 32 in this case is that a matrix in $(\ell(p), \ell_\infty)$ is also in $(c_0(p), \ell_\infty)$, though $\ell(p) \subset c_0(p)$.

(ii) If $\inf_{k \geq 0} p_k = 0$, the condition $\sup_{n,k} |a_{nk}| < \infty$ is not necessary for $A \in (\ell(p), \ell_\infty)$. To see this, let $K = \mathbb{Q}_q$ for a prime q. Consider the matrix $A \equiv (a_{nk})$ given by

$$
A \equiv \begin{bmatrix}
1 & 0 & 0 & 0 & 0 & \cdots \\
1 & q^{-1} & 0 & 0 & 0 & \cdots \\
1 & q^{-1} & q^{-2} & 0 & 0 & \cdots \\
1 & q^{-1} & q^{-2} & q^{-3} & 0 & \cdots \\
\cdots & \cdots & \cdots & \cdots & \cdots & \cdots
\end{bmatrix},
$$

$$
i.e., \ a_{nk} = q^{-k}, \ if \ 0 \leq k \leq n;
$$

$$
= 0, \ if \ k > n.
$$

Let $p_0 = 0$, $p_k = \frac{1}{k}$, $k = 1, 2, \ldots$. Then $\sup_{n,k} |a_{nk}|_q^{p_k} = \dfrac{1}{|q|_q} < \infty$, so that $A \in (\ell(p), \ell_\infty)$. However,

$$
\sup_{n,k} |a_{nk}|_q = \sup_k |q|_q^{-k} = \infty.
$$

This establishes our contention.

Bibliography

[1] J.A. Fridy. A note on absolute summability. *Proc. Amer. Math. Soc.*, 20:285–286, 1969.

[2] J.A. Fridy. Properties of absolute summability matrices. *Proc. Amer. Math. Soc.*, 24:583–585, 1970.

[3] K. Knopp and G.G. Lorentz. Beiträge Zur absoluten Limitierung. *Arch. Math.*, 2:10–16, 1949.

[4] M. Koskela. A characterization of non-negative operators on l^p to l^q with $\infty > p \geq q > 1$. *Pacific J. Math.*, 75:165–169, 1978.

[5] C.G. Lascarides and I.J. Maddox. Matrix transformations between some classes of sequences. *Proc. Cambridge Philos. Soc.*, 68:99–104, 1970.

[6] I.J. Maddox. Continuous and Köthe-Toeplitz duals of certain sequence spaces. *Proc. Cambridge Philos. Soc.*, 65:431–435, 1969.

[7] I.J. Maddox. *Elements of Functional Analysis*. Cambridge, 1977.

[8] F.M. Mears. Absolute regularity and the Nörlund mean. *Ann. of Math.*, 38:594–601, 1937.

[9] A.F. Monna. Sur le théorème de Banach-Steinhaus. *Indag. Math.*, 25:121–131, 1963.

[10] P.N. Natarajan. Multiplication of series with terms in a non-archimedean field. *Simon Stevin*, 52:157-160, 1978.

[11] P.N. Natarajan. Characterization of a class of infinite matrices with applications. *Bull. Austral. Math. Soc.*, 34:161–175, 1986.

[12] P.N. Natarajan and M.S. Rangachari. Matrix transformations between sequence spaces over non-archimedean fields. *Rev. Roum. Math. Pures Appl.*, 24:615–618, 1979.

[13] M.S. Rangachari and V.K. Srinivasan. Matrix transformations in non-archimedean fields. *Indag. Math.*, 26:422–429, 1964.

[14] M. Stieglitz and H. Tietz. Matrix transformationen von Folgenräumen eine Ergebnisübersicht. *Math. Z.*, 154: 1–16, 1977.

Chapter 4

Characterization of Regular and Schur Matrices

4.1 Introduction

When $K = \mathbb{R}$ or $\mathbb{C}$, Maddox [6] obtained a characterization of Schur matrices in terms of the existence of a bounded, divergent sequence, all of whose subsequences are summable by the matrix. This characterization is included the following earlier result of Buck [1].

Theorem 33. *A sequence $\{x_k\}$, summable by a regular matrix A, is convergent if and only if it sums each one of its subsequences.*

Maddox's result (loc. cit.) can also be stated as follows, indicating its characterizing nature.

Theorem 34. *A matrix A is a Schur matrix if and only if there exists a bounded, divergent sequence, each one of whose subsequences is summable A.*

That a Schur matrix sums all subsequences of a bounded, divergent sequence makes one part of Theorem 34 trivial. In this context, a result of Natarajan [9], characterizing the matrix class (ℓ_∞, c_0), is relevant too.

Another aspect is the study of summability of rearrangements of a bounded sequence with respect to the summability by a matrix. Fridy [3] showed that we can replace "subsequence" in Theorem 33 by "rearrangement" to obtain the following theorem.

Theorem 35. *A sequence $\{x_k\}$, summable by a regular matrix A, is convergent if and only if it sums each one of its rearrangements.*

In Section 4.2 of this chapter, we show that the analogues of Theorems 33-35 hold when K is a complete, non-trivially valued, non-Archimedean field.

In the context of rearrangements of a bounded sequence, Fridy [3] proved the following result.

Theorem 36. *A null sequence $\{x_k\}$ is in ℓ_1 if and only if there exists a matrix $A \in (\ell_1, \ell_1; P)$ which transforms all rearrangements of $\{x_k\}$ into a sequence in ℓ_1.*

Section 4.2 includes the analogue of the above result in the non-Archimedean case too.

The notion of the core of a complex sequence (see Definition 26), which was introduced by Knopp (see [2], [4]), was further studied by several authors, e.g., Sherbakoff [10]. Leaving other aspects, we can state, as follows, a characterization of a subclass of the class of regular matrices.

Theorem 37. *[2, p. 149, Theorem 6.4, II] A complex matrix $A = (a_{nk})$ is a regular matrix with*

$$\lim_{n \to \infty} \sum_{k=0}^{\infty} |a_{nk}| = 1$$

if and only if the core of any bounded sequence contains the core of its A-transform.

Remark 10. *[2, p. 140, Theorem 6.1, II]; [4, p. 55, Theorem 11] In the case of non-negative matrices, it was proved by Knopp that the core of the transform of a sequence is contained in the core of the sequence when the matrix is regular.*

In Section 4.3, it is shown (Theorem 43) that the above characterization holds for regular matrices $A = (a_{nk})$ but with the indispensable restriction

$$\overline{\lim_{n \to \infty}} \left(\sup_{k \geq 0} |a_{nk}| \right) = 1,$$

when K is a locally compact, non-trivially valued, non-Archimedean field.

4.2 Summability of subsequences and rearrangements

We begin with the analogue of Theorem 34, viz.,

Theorem 38 (see [7]). $A = (a_{nk})$, $a_{nk} \in K$, $n, k = 0, 1, 2, \ldots$, *where K is a complete, non-trivially valued, non-Archimedean field, is a Schur matrix if and only if there exists a bounded, divergent sequence $x = \{x_k\}$, each one of whose subsequences is summable A.*

Proof. When A is a Schur matrix, the assertion in the theorem is a consequence of the definition of such a matrix. Conversely, let $x = \{x_k\}$ be a bounded, divergent sequence, each one of whose subsequences is summable A. For each $p = 0, 1, 2, \ldots$, we can choose two subsequences of x (say) $\{x_k^{(1)}\}$, $\{x_k^{(2)}\}$ such that if $y_k = x_k^{(1)} - x_k^{(2)}$, $y_k = 0$, $k \neq p$, while $y_p \neq 0$. Such a choice is possible, since x diverges and so has two unequal entries after any stage for k. The sequence $\{y_k\}$ is summable A. So it follows that

$$\lim_{n \to \infty} a_{np} \text{ exists}, p = 0, 1, 2, \ldots.$$

Next we show that

$$a_{np} \to 0, \ p \to \infty, n = 0, 1, 2, \ldots.$$

For, otherwise, there exists $\epsilon' > 0$ and a non-negative integer m such that

$$|a_{m,p(i)}| > \epsilon', \ i = 1, 2, \ldots,$$

where $\{p(i)\}$ is an increasing sequence of positive integers. Since x diverges, it is not a null sequence and so there exists $\epsilon'' > 0$ and an increasing sequence $\{\ell(j)\}$ of positive integers such that

$$|x_{\ell(j)}| > \epsilon'', \ j = 1, 2, \ldots.$$

Now,

$$|a_{m,p(i)} x_{\ell(p(i))}| > \epsilon^2, \ i = 1, 2, \ldots,$$

where $\epsilon = \min(\epsilon', \epsilon'')$. This means that the A-transform of the subsequence $\{x_{\ell(j)}\}$ does not exist. Hence

$$a_{np} \to 0, p \to \infty, \ n = 0, 1, 2, \dots.$$

Now,

$$|x_{k+1} - x_k| \not\to 0, \ k \to \infty,$$

since x diverges, so that there exist $\epsilon''' > 0$ and an increasing sequence $\{k(j)\}$ of positive integers such that

$$|x_{k(j)+1} - x_{k(j)}| > \epsilon''', \ j = 1, 2, \dots. \tag{4.1}$$

We may assume that $k(j+1) - k(j) > 1, j = 1, 2, \dots.$ We claim that if A is not a Schur matrix, we should have an $\epsilon > 0$ and two strictly increasing sequences $\{n(i)\}$ and $\{p(n(i))\}$ of positive integers with

$$
\left.
\begin{array}{l}
(i) \quad \displaystyle\sup_{0 \le p \le p(n(i-1))} |a_{n(i)+1,p} - a_{n(i),p}| < \dfrac{\epsilon^2}{4M}; \\[1em]
(ii) \ |a_{n(i)+1,p(n(i))} - a_{n(i),p(n(i))}| > \epsilon; \\[1em]
and \\[1em]
(iii) \quad \displaystyle\sup_{p \ge p(n(i+1))} |a_{n(i)+1,p} - a_{n(i),p}| < \dfrac{\epsilon^2}{4M},
\end{array}
\right\} \tag{4.2}
$$

where $M = \sup_{k \ge 0} |x_k|$. Before proving the claim, we show that if A is not a Schur matrix, then x is necessarily bounded under the hypothesis of the theorem. Suppose x is unbounded. We consider two cases:

Case (i). If A is such that

$$a_{nk} \ne 0 \text{ for some } n \text{ and an infinity of } k = k(i), \ i = 1, 2, \dots,$$ choose a subsequence $\{x_{\alpha(k)}\}$ of x, which is unbounded, such that

$$|a_{nk} x_{\alpha(k)}| > 1, \ k = k(i), i = 1, 2, \dots.$$

Hence $\{x_{\alpha(k)}\}$ is not summable A, a contradiction.

Case (ii). If now, $a_{nk} = 0, k > k(n), n = 0, 1, 2, \dots, A$ not being a Schur matrix, we can find two strictly increasing sequences of positive integers $\{n(j)\}$,

$\{k(j)\}$ such that $a_{n(j),k(j)}$ is the last non-zero term in the $n(j)$th row. x, being unbounded, we can choose a subsequence $\{x_{\alpha(k)}\}$ of x such that

$$|A_{n(j)}(\{x_{\alpha(k)}\})| > j, \ j = 1, 2, \ldots.$$

To do this, choose $x_{\alpha(i)}$, $i = 1, 2, \ldots, k(1)$ such that

$$|x_{\alpha(1)}| > \frac{1}{|a_{n(1),1}|}, \quad \text{if } a_{n(1),1} \neq 0,$$

while $x_{\alpha(1)}$ is chosen as a suitable x_k otherwise. Having chosen $x_{\alpha(1)}$, choose $x_{\alpha(2)}$, $\alpha(2) > \alpha(1)$, such that

$$|x_{\alpha(2)}| > \frac{1}{|a_{n(1),2}|} \left\{ 1 + |a_{n(1),1} x_{\alpha(1)}| \right\}, \quad \text{if } a_{n(1),2} \neq 0,$$

otherwise, choose $x_{\alpha(2)}$ such that $\alpha(2) > \alpha(1)$. Now,

$$|a_{n(1),1} x_{\alpha(1)} + a_{n(1),2} x_{\alpha(2)}|$$
$$\geq |a_{n(1),2} x_{\alpha(2)}| - |a_{n(1),1} x_{\alpha(1)}|,$$
$$\text{if } a_{n(1),2} \neq 0,$$
$$= |a_{n(1),1} x_{\alpha(1)}|, \text{ if } a_{n(1),2} = 0.$$

Thus,

$$|a_{n(1),1} x_{\alpha(1)} + a_{n(1),2} x_{\alpha(2)}| = 0, \text{if } a_{n(1),1} = a_{n(1),2} = 0,$$
$$> 1, \text{if one of } a_{n(1),1}, a_{n(1),2} \text{ is not } 0.$$

Choose $x_{\alpha(k)}$, $k = 1, 2, \ldots, k(1)$ as above. Then

$$\left| \sum_{k=1}^{k(1)} a_{n(1),k} x_{\alpha(k)} \right| > 1.$$

If now, $\sum_{k=1}^{k(1)} a_{n(2),k} x_{\alpha(k)} = a$, choose similarly $x_{\alpha(k)}$, $k(1) < k \leq k(2)$ with $\alpha(k(1)) < \alpha(k(1)+1) < \cdots < \alpha(k(2))$ and

$$\left| \sum_{k=k(1)+1}^{k(2)} a_{n(2),k} x_{\alpha(k)} \right| > 2 + |a|.$$

Now,

$$\left| \sum_{k=1}^{k(2)} a_{n(2),k} x_{\alpha(k)} \right| \geq \left| \sum_{k=k(1)+1}^{k(2)} a_{n(2),k} x_{\alpha(k)} \right|$$

$$- \left| \sum_{k=1}^{k(1)} a_{n(2),k} x_{\alpha(k)} \right|$$

$$> 2 + |a| - |a|$$

$$= 2.$$

Inductively, we can therefore choose $x_{\alpha(k)}$, $k = 1, 2, \ldots$, with

$$|A_{n(j)}(\{x_{\alpha(k)}\})| = \left| \sum_{k=1}^{k(j)} a_{n(j),k} x_{\alpha(k)} \right|$$

$$> j, \; j = 1, 2, \ldots,$$

as stated earlier. It now follows that $\{x_{\alpha(k)}\}$ is not summable A, a contradiction. Thus, in both cases (i), (ii), it turns out that x is bounded, if A were not a Schur matrix.

Next, to show that $\epsilon > 0$, two increasing sequences $\{n(i)\}$ and $\{p(n(i))\}$ could be chosen to satisfy (4.2), we observe that there exist $\epsilon > 0$ and an increasing sequence $\{n(i)\}$ such that

$$\sup_{p \geq 0} |a_{n(i)+1,p} - a_{n(i),p}| > \epsilon, \; i = 1, 2, \ldots.$$

Hence there exists $p(n(i))$ such that

$$|a_{n(i)+1,p(n(i))} - a_{n(i),p(n(i))}| > \epsilon, \; i = 1, 2, \ldots. \tag{4.3}$$

Suppose $\{p(n(i))\}$ is bounded, then there are only a finite number of distinct entries in that sequence. Consequently, there exists $p = p(n(m))$ which occurs in the sequence $\{p(n(i))\}$ an infinite number of times. For this p, (4.3) will then contradict the existence of

$$\lim_{n \to \infty} a_{np} = 0, \; p = 0, 1, 2, \ldots$$

established earlier. Having chosen $\{n(i)\}$ and $\{p(n(i))\}$ to satisfy (4.3), it is clear that by choosing a subsequence of $\{n(i)\}$, if necessary, we can assume (4.2) (i), (iii) also hold along with (4.2) (ii). Consider now the sequence $\{y_p\}$ defined as follows:

$$
\left.
\begin{aligned}
y_p \;&=\; x_{k(p)+1}, \text{ if } p = p(n(i)); \\
&=\; x_{k(p)}, \text{ if } p \neq p(n(i))
\end{aligned}
\right\}, \; i = 1, 2, \ldots,
$$

with the sequence $\{k(j)\}$ already chosen as in (4.1). Thus

$$
\left| \sum_{p=0}^{\infty} \{a_{n(i)+1,p} - a_{n(i),p}\}(y_p - x_{k(p)}) \right|
$$

$$
\geq \left| \sum_{p=p(n(i-1))+1}^{p(n(i))} \{a_{n(i)+1,p} - a_{n(i),p}\}(y_p - x_{k(p)}) \right|
$$

$$
- \sum_{p=0}^{p(n(i-1))} |a_{n(i)+1,p} - a_{n(i),p}||y_p - x_{k(p)}|
$$

$$
- \sum_{p=p(n(i+1))}^{\infty} |a_{n(i)+1,p} - a_{n(i),p}||y_p - x_{k(p)}|
$$

$$
\geq |a_{n(i)+1,p(n(i))} - a_{n(i),p(n(i))}||x_{k(p(n(i)))+1} - x_{k(p(n(i)))}|
$$

$$
- \frac{\epsilon^2}{4M} M - \frac{\epsilon^2}{4M} M
$$

$$
> \epsilon^2 - \frac{\epsilon^2}{2}
$$

$$
= \frac{\epsilon^2}{2}, \; i = 1, 2, \ldots,
$$

where, we assume, as we may, that $\epsilon \geq \epsilon'''$. This is a contradiction of the fact that

$$
\left\{ \sum_{p=0}^{\infty} (a_{n+1,p} - a_{np})(y_p - x_{k(p)}) \right\}_{n=0}^{\infty}
$$

converges. This proves that

$$
\sup_{p \geq 0} |a_{n+1,p} - a_{np}| \to 0, \; n \to \infty,
$$

so that A is a Schur matrix, using Theorem 20, completing the proof of the theorem. $\qquad \square$

Remark 11. *The method of proof in Theorem 38 can be adapted to some of the theorems in the earlier chapters, for instance, Theorem 19(ii).*

Analogous to Buck's result, we have as a corollary the following:

Theorem 39. *A sequence $\{x_k\}$, $x_k \in K$, $k = 0, 1, 2, \ldots$, summable by a regular matrix A, is convergent if and only if every one of its subsequences is summable A.*

The following analogue of Theorem 35 can also be established by means of the "sliding hump method" mentioned earlier and as described by Fridy [3].

Theorem 40. *A sequence $\{x_k\}$, $x_k \in K$, $k = 0, 1, 2, \ldots$, summable by a regular matrix, is convergent if and only if every one of its rearrangements is summable A.*

There is, however, a violent deviation as regards absolute summability, viz., about sequences in ℓ_1 transformed into sequences in ℓ_1. We can combine Theorem 36 and a result of Keagy [5] to state:

Theorem 41. *When $K = \mathbb{R}$ or $\mathbb{C}$, a null sequence is in ℓ_1 if and only if there exists a matrix $A \in (\ell_1, \ell_1; P)$ which transforms every subsequence or rearrangement of x into a sequence in ℓ_1.*

The above theorem fails to hold when K is a complete, non-trivially valued, non-Archimedean field. A counterexample is provided by the matrix $A \in (\ell_1, \ell_1; P)$ defined in Remark 8(i), which sums all subsequences or rearrangements of any sequence in $c_0 - \ell_1 \neq \phi$, where $K = \mathbb{Q}_3$. However, the following result holds when K is a complete, non-trivially valued, non-Archimedean field.

Theorem 42. *A null sequence x is in ℓ_α, $\alpha > 0$, if and only if there exists a matrix $A \in (\ell_\alpha, \ell_\alpha; P)'$ such that A transforms every rearrangement of x into a sequence in ℓ_α.*

Proof. We recall that $A \in (\ell_\alpha, \ell_\alpha; P)$ if and only if

$$\sup_{k \geq 0} \sum_{n=0}^{\infty} |a_{nk}|^\alpha < \infty$$

and

$$\sum_{n=0}^{\infty} a_{nk} = 1, \ k = 0, 1, 2, \ldots.$$

Leaving the trivial part of the theorem, suppose $x \in c_0 - \ell_\alpha$ and $A \in (\ell_\alpha, \ell_\alpha; P)'$ transforms every rearrangement of x into a sequence in ℓ_α. Choose $k(1) = 1$ and a positive integer $n(1)$ such that

$$\sum_{n=n(1)+1}^{\infty} |a_{n,1}|^\alpha < 2^{-1},$$

so that

$$\sum_{n=0}^{n(1)} |a_{n,1}|^\alpha = \sum_{n=0}^{\infty} |a_{n,1}|^\alpha - \sum_{n=n(1)+1}^{\infty} |a_{n,1}|^\alpha$$

$$\geq 1 - \frac{1}{2}$$

$$= \frac{1}{2}.$$

Having defined $k(j)$, $n(j)$, $j \leq m-1$, choose a positive integer $k(m) > k(m-1)+1$ such that

$$\sum_{n=0}^{n(m-1)} |a_{n,k(m)}|^\alpha < 2^{-m},$$

$$|x_{k(m)}| < \frac{1}{m^{\frac{2}{\alpha}}}$$

and then choose a positive integer $n(m) > n(m-1)$ such that

$$\sum_{n=n(m-1)+1}^{n(m)} |a_{n,k(m)}|^\alpha \geq \frac{1}{2},$$

$$\sum_{n=n(m)+1}^{\infty} |a_{n,k(m)}|^\alpha < 2^{-m}.$$

Let U consist of all $k(m)$, $m = 1, 2, \ldots$. Let $u_m = x_{k(m)}$ and V be the set of all non-negative integers, which are not in U. Let $v = \{x_k\}_{k \in V}$. Let $y = \{y_k\}$ be the rearrangement of x, where

$$y_k = v_m, \text{ if } k = k(m);$$

$$= u_k, \text{ otherwise.}$$

Defining $n(0) = 0$, we have,

$$\sum_{n=0}^{n(M)} |(Ay)_n|^\alpha \geq \sum_{m=1}^{M} \sum_{n=n(m-1)+1}^{n(m)} \left\{ \left| \sum_{k \in U} a_{nk} y_k \right|^\alpha - \left| \sum_{k \in V} a_{nk} y_k \right|^\alpha \right\}$$

$$\geq \sum_{m=1}^{M} \sum_{n=n(m-1)+1}^{n(m)} \left\{ \left| \sum_{k \in U} a_{nk} y_k \right|^\alpha - \sum_{k \in V} |a_{nk} y_k|^\alpha \right\}$$

$$\geq \sum_{m=1}^{M} \sum_{n=n(m-1)+1}^{n(m)} \left\{ |a_{n,k(m)} v_m|^\alpha - \sum_{\substack{i=1 \\ i \neq m}}^{\infty} |a_{n,k(i)} v_i|^\alpha \right.$$

$$\left. - \sum_{k \in V} |a_{nk} y_k|^\alpha \right\}$$

$$\geq \frac{1}{2} \sum_{m=1}^{M} |v_m|^\alpha - \sum_{m=1}^{M} \sum_{n=n(m-1)+1}^{n(m)} \sum_{\substack{i=1 \\ i \neq m}}^{\infty} |a_{n,k(i)} v_i|^\alpha$$

$$- \sum_{n=0}^{n(M)} \sum_{k \in V} |a_{nk} y_k|^\alpha. \tag{4.4}$$

However,

$$\sum_{m=1}^{M} \sum_{n=n(m-1)+1}^{n(m)} \sum_{\substack{i=1 \\ i \neq m}}^{\infty} |a_{n,k(i)} v_i|^\alpha \leq \|x\|^\alpha \sum_{m=1}^{\infty} 2^{-m+1}, \tag{4.5}$$

for,

$$\sum_{m=1}^{\infty} \sum_{n=n(m-1)+1}^{n(m)} \sum_{i<m} |a_{n,k(i)}|^\alpha$$

$$= \sum_{m=1}^{\infty} \sum_{n=n(m)+1}^{\infty} |a_{n,k(m)}|^\alpha$$

$$< \sum_{m=1}^{\infty} 2^{-m},$$

and similarly

$$\sum_{m=1}^{\infty} \sum_{n=n(m-1)+1}^{n(m)} \sum_{i>m} |a_{n,k(i)}|^{\alpha} < \sum_{m=1}^{\infty} 2^{-(m+1)}.$$

Also,

$$\sum_{n=0}^{n(M)} \sum_{k \in V} |a_{nk} y_k|^{\alpha} = \sum_{k \in V} \sum_{n=0}^{n(M)} |a_{nk} u_k|^{\alpha}$$

$$\leq \left(\sup_{k \geq 0} \sum_{n=0}^{\infty} |a_{nk}|^{\alpha} \right) \left(\sum_{k \in V} |u_k|^{\alpha} \right)$$

$$< \infty. \tag{4.6}$$

In view of (4.4)-(4.6),

$$\sum_{n=0}^{n(M)} |(Ay)_n|^{\alpha} \to \infty, \ M \to \infty,$$

since $u \in \ell_{\alpha}$ and so $v \notin \ell_{\alpha}$, i.e., $Ay \notin \ell_{\alpha}$, a contradiction, proving the theorem.

□

4.3   The core of a sequence

We begin with a formal definition (see [8]).

Definition 26. *If $x = \{x_n\}$, $x_n \in K$, and $n = 0, 1, 2, \ldots$, where K is a complete, non-trivially valued, non-Archimedean field, is a sequence, we denote by $K_n(x)$, the smallest closed, K-convex set containing $x_n, x_{n+1}, \ldots$, $n = 0, 1, 2, \ldots$ and call $\mathscr{K}(x) = \bigcap_{n=0}^{\infty} K_n(x)$ the core of x.*

It is clear that a limit point of any sequence x is in $\mathscr{K}(x)$, for, if z is such a limit point, then $z = \lim_{i \to \infty} x_{n(i)}$ for some sequence $\{n(i)\}$ of positive integers and so $z \in K_p(x)$, $p = 0, 1, 2, \ldots$ and so $z \in \mathscr{K}(x) = \bigcap_{n=0}^{\infty} K_n(x)$. On the other hand, if x is unbounded, then $\mathscr{K}(x) = K$, for, $K_n(x) \neq K$, being either an absolutely K-convex set or a translate of such a set, is necessarily a closed sphere, say, $C_{r_n}(\alpha_n)$ for some $\alpha_n \in K$ and $r_n > 0$, which makes x bounded. In fact,

$$K_n(x) = C_{r_n}(x_n),$$

where

$$r_n = \sup_{k \geq n} |x_k - x_n|,$$

as is easily seen. So $K_n(x) = K$, $n = 0, 1, 2, \ldots$, if x is unbounded. Note in this context that if $\mathscr{K}(x) = K$, x is necessarily unbounded. When $K = \mathbb{R}$ or $\mathbb{C}$, this is not necessarily the case. Moreover, when K is locally compact, $\mathscr{K}(x)$ is the smallest closed, K-convex set containing the limit points of x, if x is a bounded sequence. To see this, let P be a closed, K-convex set containing the limit points of x. Then $x_n \in P$, $n \geq m$ for some m, for, otherwise, there exists a subsequence of x, all of whose elements lie outside P. Since K is locally compact and x is bounded, this sequence has a convergent subsequence, whose limit cannot be in P. Hence $K_m(x) \subset P$ and so $\mathscr{K}(x) \subset P$.

It therefore follows that if two bounded sequences have the same set of limit points, their cores are the same. There are, however, bounded sequences which have different limit points but the same core. From what has been said above, it is clear that $\mathscr{K}(x)$ is a singleton $\{y\}$ if and only if x converges to y. Generally,

$$\mathscr{K}(x) = \bigcap_{n=0}^{\infty} C_{r_n}(x_n),$$

$r_n = \sup_{k \geq n} |x_k - x_n|$. Also $\mathscr{K}(x) = C_r(\alpha)$, where α is any limit point of x and $r = \inf_{n \geq 0} r_n$.

Remark 12. *We now show (cf. Sherbakaoff [10]) that if $x = \{x_n\}$ is a bounded sequence, then*

$$\mathscr{K}(x) = \bigcap_{z \in K} C_{\overline{\lim}_{n \to \infty} |z - x_n|}(z).$$

Since any limit point y of x is clearly such that

$$|z - y| \le \varlimsup_{n \to \infty} |z - x_n|,$$

it is clear that any limit point of x is in

$$\mathscr{K}'(x) = \bigcap_{z \in K} C_{\overline{\lim}_{n \to \infty} |z - x_n|}(z).$$

Consequently,

$$\mathscr{K}(x) \subset \mathscr{K}'(x).$$

Again,

$$\varlimsup_{k \to \infty} |x_k - x_n| \le \sup_{k \ge n} |x_k - x_n|$$

so that

$$\mathscr{K}'(x) \subset \bigcap_{n=0}^{\infty} C_{\overline{\lim}_{k \to \infty} |x_k - x_n|}(x_n)$$

$$\subset \bigcap_{n=0}^{\infty} C_{\sup_{k \ge n} |x_k - x_n|}(x_n)$$

$$= \bigcap_{n=0}^{\infty} K_n(x)$$

$$= \mathscr{K}(x).$$

Following Sherbakoff [10], we can define, for $\alpha > 0$, the generalized core of a sequence $x = \{x_n\}$ by

$$\mathscr{K}^{(\alpha)}(x) = \bigcap_{z \in K} C_{\alpha \overline{\lim}_{n \to \infty} |z - x_n|}(z), \quad \text{if } x \text{ is bounded};$$

$$= K, \text{ if } x \text{ is unbounded}.$$

For $\alpha = 1$, $\mathscr{K}^{(\alpha)}(x)$ reduces to the usual core $\mathscr{K}(x)$. We now have the following analogue of Theorem 37.

Theorem 43. *An infinite matrix $A = (a_{nk})$, $a_{nk} \in K$, $n, k = 0, 1, 2, \dots$ is such that*

$$\mathscr{K}(A(x)) \subset \mathscr{K}^{(\alpha)}(x)$$

for any sequence x if and only if A is regular and satisfies

$$\varlimsup_{n \to \infty} \left(\sup_{k \geq 0} |a_{nk}| \right) \leq \alpha. \tag{4.7}$$

Proof. It suffices to consider bounded sequences. Let $x = \{x_n\}$ be a bounded sequence. If y is any limit point of $\mathscr{K}(A(x))$, then

$$|y - z| \leq \varlimsup_{n \to \infty} |z - (Ax)_n| \quad \text{for any } z \in K.$$

If A is a regular matrix satisfying (4.7),

$$|y - z| \leq \varlimsup_{n \to \infty} |z - (Ax)_n|$$

$$= \varlimsup_{n \to \infty} \left| \sum_{k=0}^{\infty} a_{nk}(z - x_k) \right|$$

$$\leq \alpha \varlimsup_{k \to \infty} |z - x_k|,$$

i.e., $y \in C_{\alpha \overline{\lim}_{k \to \infty} |z - x_k|}(z)$ for any $z \in K$.

In other words, $y \in \mathscr{K}^{(\alpha)}(x)$ and so $\mathscr{K}(A(x)) \subset \mathscr{K}^{(\alpha)}(x)$. Conversely, if $\mathscr{K}(A(x)) \subset \mathscr{K}^{(\alpha)}(x)$, then it is clear that A is regular by considering convergent sequences x, for which $\mathscr{K}^{(\alpha)}(x)$ is its limit. It remains to show that (4.7) is fulfilled. To this end, let if possible,

$$\varlimsup_{n \to \infty} \left(\sup_{k \geq 0} |a_{nk}| \right) > \alpha.$$

Illustrating Remark 11, the details, which depend on the Ganapathy Iyer-Schur technique, also called the sliding hump method, are streamlined as in the proof of Theorem 38. We can choose two strictly increasing sequences $\{n(i)\}$ and $\{k(n(i))\}$ of positive integers such that

$$\sup_{0 \leq k \leq k(n(i-1))} |a_{n(i),k}| < \alpha,$$

$$|a_{n(i),k(n(i))}| > \alpha,$$

and

$$\sup_{k \geq k(n(i+1))} |a_{n(i),k}| < \alpha.$$

Define the sequence $x = \{x_k\}$, where

$$\left. \begin{aligned} x_k &= 1, \ \text{if } k = k(n(i)) \\ &= 0, \ \text{if } k \neq k(n(i)) \end{aligned} \right\}, \ i = 1, 2, \ldots.$$

From the definition of $\mathscr{K}^{(\alpha)}(x)$, it follows that $\mathscr{K}^{(\alpha)}(x) \subset c_\alpha(0)$. However,

$$(Ax)_{n(i)} = \sum_{k=0}^{k(n(i-1))} a_{n(i),k} x_k + \sum_{k=k(n(i-1))+1}^{k(n(i))} a_{n(i),k} x_k$$

$$+ \sum_{k=k(n(i+1))}^{\infty} a_{n(i),k} x_k$$

$$= \sum_{k=0}^{k(n(i-1))} a_{n(i),k} x_k + a_{n(i),k(n(i))}$$

$$+ \sum_{k=k(n(i+1))}^{\infty} a_{n(i),k} x_k.$$

So,

$$\alpha < |a_{n(i),k(n(i))}|$$

$$\leq \max \left\{ |(Ax)_{n(i)}|, \sup_{0 \leq k \leq k(n(i-1))} |a_{n(i),k}|, \right.$$

$$\left. \sup_{k \geq k(n(i+1))} |a_{n(i),k}| \right\}$$

$$< \max\{|(Ax)_{n(i)}|, \alpha, \alpha\}.$$

Consequently,

$$|(Ax)_{n(i)}| > \alpha, \ i = 1, 2, \ldots. \tag{4.8}$$

Since $A \in (\ell_\infty, \ell_\infty)$, by the regularity of A, $\{(Ax)_{n(i)}\}_{i=1}^{\infty} \in \ell_\infty$. Since K is locally compact, it has a convergent subsequence, whose limit cannot be in $c_\alpha(0)$ because of (4.8), contradicting the fact $\mathscr{K}(A(x)) \subset \mathscr{K}^{(\alpha)}(x)$. Hence

$$\varlimsup_{n \to \infty} \left(\sup_{k \geq 0} |a_{nk}| \right) \leq \alpha,$$

completing the proof of the theorem. $\qquad \square$

Remark 13. *(i) When $K = \mathbb{R}$ or $\mathbb{C}$, Sherbakoff [10] has shown that under the condition*

$$\overline{\lim_{n \to \infty}} \left(\sum_{k=0}^{\infty} |a_{nk}| \right) = \alpha, \ \alpha \geq 1, \tag{4.9}$$

$\mathcal{K}(A(x)) \subset \mathcal{K}^{(\alpha)}(x)$. His proof, however, works with the less stringent condition

$$\overline{\lim_{n \to \infty}} \left(\sum_{k=0}^{\infty} |a_{nk}| \right) \leq \alpha, \ \alpha \geq 1. \tag{4.10}$$

It can also be shown, as in the proof of Theorem 43, that condition (4.10) is necessary in addition to regularity of the matrix for $\mathcal{K}(A(x)) \subset \mathcal{K}^{(\alpha)}(x)$ for any bounded sequence x. The details of the proof are left to the reader, being analogous to a part of those in Theorem 43 (see [8]).

(ii) Condition (4.7) or its analogous condition (4.10) cannot be relaxed if $\mathcal{K}(A(x))$ were to be contained in $\mathcal{K}^{(\alpha)}(x)$. In any case, the matrix

$$\begin{bmatrix} 1 & \lambda & -\lambda & 0 & 0 & 0 & \dots \\ 0 & 1 & \lambda & -\lambda & 0 & 0 & \dots \\ 0 & 0 & 1 & \lambda & -\lambda & 0 & \dots \\ \dots & \dots & \dots & \dots & \dots & \dots & \dots \end{bmatrix},$$

where $|\lambda| > \alpha$, transforms the sequence $\{0, 1, 0, 1, \dots\}$ into the sequence $\{\lambda, 1 - \lambda, \lambda, 1 - \lambda, \dots\}$. $\mathcal{K}^{(\alpha)}(x) \subset C_{\alpha}(0)$, while $\lambda \in \mathcal{K}(A(x))$ and $\lambda \notin C_{\alpha}(0)$.

(iii) For a regular matrix $A = (a_{nk})$,

$$\overline{\lim_{n \to \infty}} \sum_{k=0}^{\infty} |a_{nk}| \leq 1$$

if and only if

$$\lim_{n \to \infty} \sum_{k=0}^{\infty} |a_{nk}| = 1.$$

Bibliography

[1] R.C. Buck. A note on subsequences. *Bull. Amer. Math. Soc.*, 49:898–899, 1943.

[2] R.G. Cooke. *Infinite matrices and sequence spaces*. Macmillan, 1950.

[3] J.A. Fridy. Summability of rearrangements of sequences. *Math. Z.*, 143:187–192, 1975.

[4] G.H. Hardy. *Divergent Series*. Oxford, 1949.

[5] T.A. Keagy. Matrix transformations and absolute summability. *Pacific J. Math.*, 63:411–415, 1976.

[6] I.J. Maddox. A Tauberian theorem for subsequences. *Bull. London Math. Soc.*, 2:63–65, 1970.

[7] P.N. Natarajan. *Characterization of regular and Schur matrices over non-archimedean fields. Proc. Kon. Ned. Akad. Wetens, Series A*, 90:423–430, 1987.

[8] P.N. Natarajan. On the core of a sequence over valued fields. *J. Indian. Math. Soc.*, 55:189–198, 1990.

[9] P.N. Natarajan. A characterization of the matrix class (l_∞, c_0). *Bull. London Math. Soc.*, 23:267–268, 1991.

[10] A.A. Sherbakoff. On cores of complex sequences and their regular transforms (Russian). *Mat. Zametki*, 22:815–823, 1977.

Chapter 5

A Study of the Sequence Space $c_0(p)$

5.1 Identity of weak and strong convergence or the Schur property

In the present chapter, we are concerned with sequence spaces over a complete, non-trivially valued field K, Archimedean or non-Archimedean. It will be stated explicitly in the relevant context when the field is Archimedean or non-Archimedean. More particularly, we study the family $c_0(p)$ of sequence spaces defined in Chapter 1 (see Definition 20). Unlike the earlier chapters, we write the sequence $\{x_k\}$, beginning with $k = 1$, for the sake of convenience.

For the contents of this chapter, one can refer to [1].

The topology on $c_0(p)$, with which we will be concerned in what follows, is given by the metric

$$d(x, y) = g(x - y),$$

where g, defined by (1.7), is a paranorm in the sense of Maddox [7]. $c_0(p)$ is then a complete metric linear space and hence a locally convex (locally K-convex [20]) space when $K = \mathbb{R}$ or $\mathbb{C}$ (a complete, non-trivially valued, non-Archimedean field). This topology on $c_0(p)$ can also be described in terms of the seminorms

$$\wp_\alpha(x) = \sup_{k \geq 1} |x_k| \alpha^{\frac{1}{p_k}}, \ \ \alpha = 1, 2, \dots, x = \{x_k\} \in c_0(p). \tag{5.1}$$

To see this, we first observe that $c_0(p)$ is complete under the seminorms in (5.1). If $\{x^{(n)}\}$ is a Cauchy sequence in $c_0(p)$ under the seminorms in (5.1), $x^{(n)} = \{x_k^{(n)}\}_{k=1}^{\infty}$, $n = 1, 2, \ldots$, then

$$\wp_\alpha(x^{(m)} - x^{(n)}) \to 0, \ m, n \to \infty, \alpha = 1, 2, \ldots,$$

$$i.e., \sup_{k \geq 1} |x_k^{(m)} - x_k^{(n)}| \alpha^{\frac{1}{p_k}} \to 0, \ m, n \to \infty, \alpha = 1, 2, \ldots. \tag{5.2}$$

(5.2) is equivalent to

$$\sup_{k \geq 1} |x_k^{(m)} - x_k^{(n)}|^{p_k} \to 0, \ m, n \to \infty. \tag{5.3}$$

To prove this equivalence, let (5.2) hold while (5.3) does not. Then for some $\epsilon > 0$ and subsequences $\{m(j)\}_{j=1}^{\infty}$ and $\{n(i)\}_{i=1}^{\infty}$ of positive integers

$$\sup_{k \geq 1} |x_k^{(m(j))} - x_k^{(n(i))}|^{p_k} > \epsilon, \ i, j = 1, 2, \ldots.$$

If the positive integer α_0 is such that $\frac{1}{\alpha_0} < \epsilon$, then

$$\sup_{k \geq 1} |x_k^{(m(j))} - x_k^{(n(i))}|^{p_k} > \frac{1}{\alpha_0}, \ i, j = 1, 2, \ldots.$$

So

$$\sup_{k \geq 1} |x_k^{(m(j))} - x_k^{(n(i))}| \alpha_0^{\frac{1}{p_k}} > 1, \ i, j = 1, 2, \ldots,$$

contradicting (5.2). Thus (5.2) implies (5.3). To prove the reverse implication, suppose (5.3) holds while (5.2) does not. Then there exists ϵ with $0 < \epsilon < 1$, an integer $\alpha_0 \geq 1$, subsequences $\{m(j)\}_{j=1}^{\infty}$ and $\{n(i)\}_{i=1}^{\infty}$ of positive integers such that

$$\sup_{k \geq 1} |x_k^{(m(j))} - x_k^{(n(i))}| \alpha_0^{\frac{1}{p_k}} > \epsilon, \ i, j = 1, 2, \ldots,$$

$$i.e., \sup_{k \geq 1} |x_k^{(m(j))} - x_k^{(n(i))}|^{p_k} > \frac{\epsilon^H}{\alpha_0}, \ i, j = 1, 2, \ldots.$$

This contradicts (5.3). What we have observed now is the fact that $c_0(p)$ has the same set of Cauchy sequences under the seminorms in (5.1) and under the paranorm given by (1.7). Further, if $\{x^{(n)}\}$ is a Cauchy sequence in $c_0(p)$ under the seminorms in (5.1), then (5.3) holds. By the completeness of $c_0(p)$ under the paranorm (1.7), it follows that

$$x^{(n)} \to x, \ n \to \infty, \text{ where } x^{(n)} = \{x_k^{(n)}\}_{k=1}^{\infty}, x = \{x_k\}_{k=1}^{\infty} \in c_0(p).$$

Thus the sets of convergent sequences in the paranorm (1.7) and the seminorms in (5.1) are one and the same with the limits of these sequences also being the same. Thus the topology given by the paranorm (1.7) is the same as the locally convex (locally K-convex) topology given by the seminorms in (5.1). It is easily verified that the continuous dual $c_0^*(p)$ of $c_0(p)$ consists of functionals f:

$$
\left.
\begin{aligned}
&(i) \ f(x) = \sum_{k=1}^{\infty} a_k x_k, \{a_k\} \in M_0(p), \text{ when } K = \mathbb{R} \text{ or } \mathbb{C}[7]; \\
&(ii) \ f(x) = \sum_{k=1}^{\infty} a_k x_k, \{a_k\} \in l_\infty(p), \text{ when } K \text{ is a complete,} \\
&\qquad \text{non-trivially valued, non-Archimedean field.}
\end{aligned}
\right\} \quad (5.4)
$$

The topology in $c_0^*(p)$ is given by the following scheme of convergence: $f_n \to f$, $n \to \infty$, $f_n, f \in c_0^*(p)$ means $f_n(x) \to f(x)$, $n \to \infty$, uniformly with respect to x in any sphere around the origin.

The following result, which characterizes the matrix class $(l_\infty(p), c_0)$, helps us to prove that weak and strong convergence are equivalent in $c_0(p)$, i.e., $c_0(p)$ has the Schur property, when K is a complete, non-trivially valued, non-Archimedean field. The case $K = \mathbb{R}$ or $\mathbb{C}$ was already dealt with by Lascarides [6]. The result, however, asserts the same conclusion if and only if

$$
\lim_{k \to \infty} p_k = 0.
$$

Lemma 1. *If K is a complete, non-trivially valued, non-Archimedean field, $A = (a_{nk}) \in (\ell_\infty(p), c_0)$ if and only if*

(i) $|a_{nk}|^{p_k} \to 0, k \to \infty, n = 1, 2, \ldots$,

    i.e., $\{a_{nk}\}_{k=1}^{\infty} \in c_0(p), n = 1, 2, \ldots$;

    and

(ii) $\sup_{k \geq 1} |a_{nk}| N^{\frac{1}{p_k}} \to 0, \ n \to \infty$ *for every integer $N > 1$.*

Proof. Sufficiency. Let (i), (ii) hold. Then

$$
\{a_{nk}\}_{k=1}^{\infty} \in c_0(p) = \ell_\infty^{\times}(p), n = 1, 2, \ldots.
$$

If $x = \{x_k\} \in \ell_\infty(p)$ and $|x_k|^{p_k} < M$, M being an integer > 1, $k = 1, 2, \ldots$, $\sum_{k=1}^{\infty} a_{nk} x_k$ converges, $n = 1, 2, \ldots$. Also,

$$|(Ax)_n| = \left| \sum_{k=1}^{\infty} a_{nk} x_k \right|$$

$$\leq \sup_{k \geq 1} |a_{nk}| |x_k|$$

$$\leq \sup_{k \geq 1} |a_{nk}| M^{\frac{1}{p_k}}$$

$$\to 0, n \to \infty,$$

$$i.e., \ \{(Ax)_n\} \in c_0.$$

Necessity. Let $\{(Ax)_n\} \in c_0$, whenever $x = \{x_k\} \in \ell_\infty(p)$. Then $\{a_{nk}\}_{k=1}^{\infty} \in \ell_\infty^\times(p) = c_0(p)$, $n = 1, 2, \ldots$. Thus (i) holds. If (ii) does not hold, then for some integer $N > 1$,

$$\sup_{k \geq 1} |a_{nk}| N^{\frac{1}{p_k}} \not\to 0, n \to \infty.$$

We observe that for every $k = 1, 2, \ldots$, there exists an integer α_k such that

$$\rho^{\alpha_k+1} \leq N^{\frac{1}{p_k}} < \rho^{\alpha_k},$$

where $0 < \rho = |\pi| < 1$, $\pi \in K$. Consider the sequence $x = \{x_k\}$, where

$$x_k = \pi^{\alpha_k}, k = 1, 2, \ldots.$$

Then

$$\sup_{k \geq 1} |a_{nk}| |\pi|^{\alpha_k} > \sup_{k \geq 1} |a_{nk}| N^{\frac{1}{p_k}}$$

$$\not\to 0, n \to \infty.$$

Thus the matrix $B = (a_{nk} \pi^{\alpha_k})$, $n, k = 1, 2, \ldots$ does not transform bounded sequences into null sequences [10]. Hence there exists $x = \{x_k\} \in \ell_\infty$ such that

$$\left\{ \sum_{k=1}^{\infty} a_{nk} \pi^{\alpha_k} x_k \right\}_{n=1}^{\infty} \notin c_0.$$

Let $y = \{y_k\}$, where $y_k = \pi^{\alpha_k} x_k$, $k = 1, 2, \ldots$.

$|y_k|^{p_k} = \rho^{\alpha_k p_k} |x_k|^{p_k} \leq L\rho^{\alpha_k \rho_k} \leq LN\rho^{-M}$, where $L = \sup\limits_{k \geq 1} |x_k|^{p_k}$, $M = \sup\limits_{k \geq 1} p_k$. Thus $y = \{y_k\} \in \ell_\infty(p)$, while $\{(Ay)_n\} \notin c_0$, which is a contradiction. Thus (ii) holds, completing the proof. $\qquad\square$

Theorem 44. *If K is a complete, non-trivially valued, non-Archimedean field, weak and strong convergence in $c_0(p)$ are equivalent, i.e., $c_0(p)$ has the Schur property.*

Proof. It suffices to show that weak convergence implies strong convergence in $c_0(p)$. Let

$$y^{(n)} \to y, n \to \infty \text{ weakly in } c_0(p),$$

where $y^{(n)} = \{y_k^{(n)}\}_{k=1}^\infty$, $n = 1, 2, \ldots$, $y = \{y_k\} \in c_0(p)$. This means that

$$f(y^{(n)} - y) \to 0, n \to \infty \text{ for every } f \in c_0^*(p).$$

By the facts mentioned in Section 1.6, $f \in c_0^*(p)$ if and only if there exists $x = \{x_k\} \in \ell_\infty(p)$ such that $f(z) = \sum\limits_{k=1}^\infty z_k x_k$, $z = \{z_k\} \in c_0(p)$. Now,

$$f(y^{(n)} - y) = \sum_{k=1}^\infty (y_k^{(n)} - y_k) x_k = \sum_{k=1}^\infty b_{nk} x_k,$$

where $b_{nk} = y_k^{(n)} - y_k$, $n, k = 1, 2, \ldots$. Hence, if $B = (b_{nk})$, $n, k = 1, 2, \ldots$, $Bx \in c_0$ for every $x \in \ell_\infty(p)$. By Lemma 1, it follows that

$$\sup_{k \geq 1} |b_{nk}| N^{\frac{1}{p_k}} \to 0, n \to \infty \text{ for every integer } N > 1.$$

This means that

$$\sup_{k \geq 1} |b_{nk}|^{p_k} \to 0, n \to \infty$$

or equivalently,

$$\{g(y^{(n)} - y)\}^H = \sup_{k \geq 1} |y_k^{(n)} - y_k|^{p_k}$$

$$= \sup_{k \geq 1} |b_{nk}|^{p_k}$$

$$\to 0, n \to \infty,$$

which implies that

$$g(y^{(n)} - y) \to 0, n \to \infty,$$

$$i.e., y^{(n)} \to y, n \to \infty$$

in the paranorm of $c_0(p)$, completing the proof of the theorem. $\qquad\square$

5.2   Normability

In contrast to the coincidence of weak and strong convergence, the necessary and sufficient condition for the metric linear space $c_0(p)$ to be normable turns out to be

$$\inf_{k \geq 1} p_k > 0,$$

regardless of the complete, non-trivially valued field K. The case $K = \mathbb{R}$ or $\mathbb{C}$ has been dealt with by Maddox and Roles [8]. The proof of Theorem 45 below, is applicable to any complete field K with non-trivial valuation.

Theorem 45. *When K is a complete, non-trivially valued field, $c_0(p)$ is normable if and only if*

$$\inf_{k \geq 1} p_k > 0$$

and in such a case, $c_0(p) = c_0$.

Proof. If $\inf_{k \geq 1} p_k = 0$, $U(\epsilon) = \{\alpha \in c_0(p) : g(\alpha) < \epsilon\}$, where g is the paranorm on $c_0(p)$, we claim that $U(\epsilon)$ is not bounded, i.e., given $U(\epsilon)$, there exists $U(\eta)$ with no $\lambda \in K$, $\lambda \neq 0$ such that $U(\epsilon) \subset \lambda U(\eta)$. Let $\lambda \in K$, $\lambda \neq 0$. There exists a subsequence $\{k(i)\}$ of positive integers such that $\lim\limits_{i \to \infty} \dfrac{p_{k(i)}}{H} = 0$, $H = \max(1, \sup\limits_{k \geq 1} p_k)$ so that $|\lambda|^{\frac{p_{k(i)}}{H}} \to 1$, $i \to \infty$. Thus for some $k(m)$, $|\lambda|^{\frac{p_{k(m)}}{H}} < 2$. Since K is non-trivially valued, there exists $\pi \in K$ such that $0 < \rho = |\pi| < 1$. Choose a positive integer m_2 (which, of course, depends on $k(m)$) such that

$$\rho^{m_2+1} \leq \left(\frac{\epsilon}{2}\right)^{\frac{H}{p_{k(m)}}} < \rho^{m_2}.$$

Let $\alpha = \{0, 0, \ldots, 0, \pi^{m_2+1}, 0, \ldots\}$, where all the entries except the $k(m)$th are zero. Then $\alpha \in c_0(p)$, while

$$g(\alpha) = \rho^{(m_2+1)\frac{p_{k(m)}}{H}}$$

$$\leq \frac{\epsilon}{2}$$

$$< \epsilon,$$

which implies that $\alpha \in U(\epsilon)$. However,

$$g\left(\frac{\alpha}{\lambda}\right) = \frac{\rho^{(m_2+1)\frac{p_{k(m)}}{H}}}{|\lambda|^{\frac{p_{k(m)}}{H}}}$$

$$> \frac{\epsilon}{4}\rho^{\frac{p_{k(m)}}{H}}$$

$$\geq \frac{\epsilon}{4}\rho,$$

since $0 < \rho < 1$ and $\frac{p_{k(m)}}{H} \leq 1$. Taking $\eta = \frac{\epsilon\rho}{4}$, we see that $\alpha \notin \lambda U(\eta)$. Thus $U(\epsilon) \not\subset \lambda U(\eta)$. Hence if $c_0(p)$ is normable, $\inf\limits_{k \geq 1} p_k > 0$.

Conversely, if $\inf\limits_{k \geq 1} p_k = p > 0$, consider $U(1)$. Take any $U(\eta)$. Choose $\mu \in K$ such that

$$|\mu| > \max\left(1, \frac{1}{\eta^{\frac{H}{p}}}\right).$$

Let $x \in U(1)$. Now,

$$g\left(\frac{x}{\mu}\right) = \sup_{k \geq 1}\left|\frac{x_k}{\mu}\right|^{\frac{p_k}{H}}$$

$$= g(x)\sup_{k \geq 1}\frac{1}{|\mu|^{\frac{p_k}{H}}}$$

$$\leq \frac{1}{|\mu|^{\frac{p}{H}}}$$

$$< \eta.$$

This implies that $\frac{x}{\mu} \in U(\eta)$ or equivalently, $x \in \mu U(\eta)$. Hence $U(1) \subset \mu U(\eta)$. $U(1)$ is thus a bounded convex (K-convex) neighborhood of 0. By the Kolmogoroff criterion ([5, p. 160, Section 15, 10(4)], [9]) for normability, the theorem follows from what we have shown and from the observation made in Section 1.6. This completes the proof of the theorem. □

Remark 14. *$c_0(p) = c_0$ if and only if $c_0(p)$ and c_0 are topologically isomorphic to each other. That they are topologically isomorphic when $c_0(p) = c_0$ can be verified by showing that the sets of convergent sequences for both these topologies are one and the same (see Section 5.1 for the method of procedure). The interesting aspect is that there cannot be a proper subspace of c_0 of the form $c_0(p)$ which is topologically isomorphic to c_0.*

5.3   Nuclearity of $c_0(p)$

In this section, $K = \mathbb{R}$ or $\mathbb{C}$. We begin with some definitions. The linear spaces which we consider in this section are Fréchet spaces in the sense of Köthe ([5, p. 164]). They are complete, metrizable locally convex spaces whose topology is given by a sequence $\{\wp_k\}$ of seminorms such that $\wp_1 \leq \wp_2 \leq \ldots$. For instance, $c_0(p)$ is a Fréchet space with the sequence of seminorms defined by (5.1) in Section 5.1. In fact, each $\wp_\alpha$ is a norm.

Definition 27. *The Fréchet space F with the increasing sequence $\{\wp_\alpha\}$ of seminorms is a "nuclear space" if given β, there exists $\alpha > \beta$ such that the canonical mapping*

$$\left(F \big/ \wp_\alpha^{-1}(0)\right)^{\sim} \to \left(F \big/ \wp_\beta^{-1}(0)\right)^{\sim}$$

obtained by extending the mapping

$$x + \wp_\alpha^{-1}(0) \to x + \wp_\beta^{-1}(0),$$

is nuclear as between two Banach spaces, where "$\sim$" stands for the completion of the normed linear space in question.

It may be recalled that a continuous linear mapping T from a normed linear space E to a Banach space F is called nuclear if there are elements

$a_n \in E^*$ and $y_n \in F$, $n = 1, 2, \ldots$ with $\sum_{n=1}^{\infty} \|a_n\| \|y_n\| < \infty$ such that T has

the form $\sum_{n=1}^{\infty} a_n(x) y_n$ for $x \in E$.

Most of the familiar sequence spaces such as c_0, ℓ_∞ are not nuclear since they are Banach spaces of infinite dimension (see [12, p. 84, Theorem 4.4.14]). Thus it is interesting to seek nuclear spaces among sequence spaces. In view of Theorem 45, $c_0(p)$ is not nuclear when $\inf_{k \geq 1} p_k > 0$, for, it becomes normable and it is of infinite dimension too ([12, loc. cit.]). Thus the condition

$$\inf_{k \geq 1} p_k = 0 \tag{5.5}$$

is necessary for $c_0(p)$ to be nuclear.

Schaefer ([15, p. 107, Section 3.8(4)]) remarks that $c_0(p)$, where $p_k = \frac{1}{k}$, $k = 1, 2, \ldots$, namely, the space of entire functions is a nuclear space. In the context of this remark, it is worthwhile investigating necessary and sufficient conditions on the sequence $\{p_k\}$ which make $c_0(p)$ nuclear. It is convenient to formulate this problem in terms of power series spaces.

Let P be a set of real sequences possessing the following properties:

(i) If $\{x_k\} \in P$, $x_k \geq 0$;

(ii) corresponding to each positive integer n, there exists a sequence $\{x_k\} \in P$ with $x_n > 0$;

and

(iii) if $\{x_k^{(1)}\}, \ldots, \{x_k^{(n)}\}$ is a finite number of sequences in P, there exists a sequence $\{x_k\} \in P$ such that

$$\max(x_k^{(1)}, \ldots, x_k^{(n)}) \leq x_k, \quad k = 1, 2, \ldots.$$

Under these assumptions, $P^\times$, the Köthe-Toeplitz dual of P (vide. Definition 21), whose elements are complex sequences, is a complete, locally convex, topological linear space with termwise addition and scalar multiplication of sequences. The seminorms on the space are given by

$$\rho_\alpha(\{x_k\}) = \sum_{k=1}^{\infty} |x_k|\alpha_k, \ \alpha = \{\alpha_k\} \in P, \{x_k\} \in P^\times.$$

It is interesting to note in this context that $c_0 \neq P^\times$ for any P. To see this, we write

$$c_0^+ = \{\{x_k\} \in c_0, x_k \geq 0, k = 1, 2, \dots\},$$

$$\ell_\infty^+ = \{\{x_k\} \in \ell_\infty, x_k \geq 0, k = 1, 2, \dots\},$$

$$\ell_1^+ = \{\{x_k\} \in \ell_1, x_k \geq 0, k = 1, 2, \dots\},$$

and note that $(c_0^+)^\times = \ell_1$, $(\ell_\infty^+)^\times = \ell_1$, $(\ell_1^+)^\times = \ell_\infty$. We also note that $P \subset Q$ implies that $P^\times \supset Q^\times$. Thus, if $c_0 = P^\times$ for some P, then $\ell_1^+ \subset P \subset \ell_\infty^+$. P cannot be ℓ_1^+ (ℓ_∞^+), since in this case $P^\times = \ell_\infty$ (ℓ_1). If $\ell_1^+ \subsetneq P$, there exists $\{z_k\} \in P$, $\{z_k\} \notin \ell_1^+$. But

$$P^\times = \left\{ \{y_k\} : \sum_{k=1}^{\infty} |y_k||x_k| < \infty \ \text{for every} \ \{x_k\} \in P \right\}.$$

Hence, in particular, $\sum_{k=1}^{\infty} y_k z_k < \infty$ for every $\{y_k\} \in c_0^+$, which means that $\{z_k\} \in (c_0^+)^\times = \ell_1$; in fact, $\{z_k\} \in \ell_1^+$, a contradiction. This contradiction establishes the claim made above.

The following theorem gives a necessary and sufficient condition for the nuclearity of $P^\times$.

Theorem 46. *[12, p. 98, Theorem 6.1.2] The locally convex space $P^\times$ is nuclear if and only if for each sequence $\{x_k\} \in P$, there are sequences $\{y_k\} \in P$, $\{z_k\} \in \ell_1$ such that*

$$x_k \leq y_k z_k, \ k = 1, 2, \dots.$$

As a consequence, the locally convex space $P^\times$ is nuclear if and only if its topology is given by the seminorms

$$\rho'_\alpha(\{x_k\}) = \sup_{k \geq 1} |x_k|\alpha_k, \ \alpha = \{\alpha_k\} \in P, \{x_k\} \in P^\times.$$

In view of Theorem 46 above, the problem about nuclearity of $c_0(p)$ is related to its being expressed as $P^\times$ for some P. It is more convenient to examine whether it is a power series space in the following sense:

Definition 28. *A space* S *of complex sequences is said to be a "power series space," if* $S = P^\times$, *where* P *consists of sequences of the form* $\{x^{\alpha_k}\}$ *for* $0 < x < x_0$ *(x_0 fixed and may be $+\infty$) and* $\{\alpha_k\}$ *is a non-decreasing sequence of non-negative numbers.*

The nomenclature is obvious, because $P^\times$ is the space of all power series with radius of convergence not less than x_0, when $\alpha_k = k$.

Theorem 46 gives the following criterion for a power series space to be nuclear.

Theorem 47. *A power series space is nuclear if and only if*

$$\sum_{k=1}^{\infty} q^{\alpha_k} < \infty \quad \text{for every } q \text{ with } 0 < q < 1$$

($\sum_{k=1}^{\infty} q^{\alpha_k} < \infty$ for some q with $0 < q < 1$), if $x_0 < +\infty$ ($x_0 = +\infty$).

Note 48. *Not all power series spaces are nuclear. For instance, take* $\alpha_k = (\log k)^r$, $0 < r < 1$. *In this case,* $\sum_{k=1}^{\infty} q_k = \infty$ *for every q with $0 < q < 1$.*

$c_0(p)$, $p_k = \frac{1}{k}$, $k = 1, 2, \dots$ is obviously a power series space, corresponding to $x_0 = +\infty$ and $\alpha_k = k$ and is nuclear (as remarked earlier) since $\sum_{k=1}^{\infty} q^{\alpha_k} < \infty$ for any q with $0 < q < 1$.

Note 49. *In this section, we do not assume that* $\{\alpha_k\}$ *is a non-decreasing sequence of non-negative numbers as this assumption does not seem to be necessary to establish the results stated or obtained in this section.*

In the context of this chapter, we do not make any distinction between power series spaces of finite and infinite type. The following lemma is very helpful in our study.

Lemma 2. *If there exists q with $0 < q < 1$ such that $\displaystyle\sum_{k=1}^{\infty} q^{\frac{1}{p_k}} < \infty$, then the following statements are equivalent:*

(i) $\displaystyle\sum_{k=1}^{\infty} |x_k| n^{\frac{1}{p_k}} < \infty$, $n = 1, 2, \ldots$;

   and

(ii) $|x_k|^{p_k} \to 0$, $k \to \infty$.

Proof. Suppose (i) holds. Then

$$|x_k| n^{\frac{1}{p_k}} \to 0, \ k \to \infty, n = 1, 2, \ldots.$$

If possible, let (ii) fail to hold. Then there exist $\epsilon > 0$ and a subsequence $\{k(i)\}$ of positive integers such that

$$|x_{k(i)}|^{p_{k(i)}} > \epsilon, \ i = 1, 2, \ldots.$$

Choose a positive integer n such that $\frac{1}{n} < \epsilon$. Then

$$|x_{k(i)}| n^{\frac{1}{p_{k(i)}}} > 1, \ i = 1, 2, \ldots,$$

which contradicts (i). Thus (i) implies (ii). Let, now, (ii) hold. By hypothesis, there exists q with $0 < q < 1$ such that $\displaystyle\sum_{k=1}^{\infty} q^{\frac{1}{p_k}} < \infty$. Given any integer $n \geq 1$, let $\epsilon = \frac{q}{n}$. By (ii), corresponding to ϵ, there exists a positive integer N such that

$$|x_k|^{p_k} < \epsilon, \ k \geq N.$$

So

$$\sum_{k=N}^{\infty} |x_k| n^{\frac{1}{p_k}} < \sum_{k=N}^{\infty} (\epsilon n)^{\frac{1}{p_k}}$$

$$= \sum_{k=N}^{\infty} q^{\frac{1}{p_k}}$$

$$< \infty,$$

i.e., (i) holds. Hence (ii) implies (i), completing the proof of the lemma. □

Lemma 3. *[6] $c_0(p) \subset \ell_1$ if and only if there exists q with $0 < q < 1$ such that*

$$\sum_{k=1}^{\infty} q^{\frac{1}{p_k}} < \infty.$$

The following results in this section are in the nature of sufficient conditions for $c_0(p)$ to be nuclear.

Theorem 50. *$c_0(p)$ is nuclear if there exists q with $0 < q < 1$ such that*
$$\sum_{k=1}^{\infty} q^{\frac{1}{p_k}} < \infty.$$

Proof. By Lemma 2, $c_0(p)$ is the space S of all sequences $\{x_k\}$ such that

$$\sum_{k=1}^{\infty} |x_k| n^{\frac{1}{p_k}} < \infty, \ n = 1, 2, \ldots.$$

The latter space is a power series space according to Definition 28. It is further nuclear in view of Theorem 47. Thus the topology of the power series space S, by Theorem 46, is given by the seminorms in (5.1). We have already shown in Section 5.1 that the topology of the seminorms in (5.1) is equivalent to the topology of the paranorm given by (1.7). Thus $c_0(p)$ and the power series space S are one and the same topologically too. Consequently $c_0(p)$ is nuclear because of Theorem 47. $\qquad \square$

The following theorem yields a sufficient condition, in another form, for the nuclearity of $c_0(p)$ in its corollary stated after its proof.

Theorem 51. *The following statements are equivalent:*

(i) *The power series space S corresponding to $x_0 = +\infty$ and $\alpha_k = \frac{1}{p_k}$ is the same as $c_0(p)$ as sets and topologically too;*

(ii) *there exists q with $0 < q < 1$ such that $\sum_{k=1}^{\infty} q^{\frac{1}{p_k}} < \infty$;*

and

(iii) *$c_0(p) \subset \ell_1$.*

Proof. The equivalence of (ii) and (iii) is just Lemma 3. (ii) implies (i) is precisely Theorem 50. If (i) holds, the topology of S is given by the seminorms in (5.1). So, by Theorem 46, S is nuclear. (ii) now follows from Theorem 47, completing the proof. $\square$

Corollary 3. *If any one of the conditions in Theorem 51 holds, then $c_0(p)$ is nuclear.*

Remark 15. *(1) There are sequences $p = \{p_k\}$ for which $c_0(p) \not\subset \ell_1$. For instance, $p_k = \frac{1}{(\log k)^r}$, $0 < r < 1$, $k = 2, 3, \ldots$. On the other hand, if $p_k = O(\frac{1}{k^r})$, $0 < r \leq 1$, then $c_0(p) \subset \ell_1$.*

(2) When (ii) in Theorem 51 is satisfied, $\lim\limits_{k \to \infty} p_k = 0$.

5.4 $c_0(p)$ as a Schwartz space

Throughout this section, K denotes a complete, non-trivially valued, non-Archimedean field. In this section, we will assume that the locally K-convex spaces considered are non-Archimedean metrizable and, therefore, that their topologies are given by increasing sequences of non-Archimedean seminorms. Besides the well-known spaces c_0, ℓ_∞ which are non-Archimedean Banach spaces and so locally K-convex spaces, $c_0(p)$ is a complete, locally K-convex space, whose topology is given by the non-Archimedean seminorms in (5.1).

In this section, E is a locally K-convex topological linear space whose non-Archimedean seminorms form an increasing sequence.

We recall that ([19, p. 134]) a subset A of a non-Archimedean normed linear space E is a compactoid if for every $\epsilon > 0$, there exists a finite set $\{x_1, x_2, \ldots, x_n\}$ of points in E such that

$$A \subset B_\epsilon(0) + \overline{co}\{x_1, x_2, \ldots, x_n\},$$

where $\overline{co}\{x_1, x_2, \ldots, x_n\}$ is the closed absolute K-convex hull of $\{x_1, x_2, \ldots, x_n\}$ and $B_\epsilon(0) = \{x \in E : \|x\| \leq \epsilon\}$.

The following definition is needed to define Schwartz spaces.

Definition 29. *Let E be a non-Archimedean normed linear space and F be a non-Archimedean Banach space. A linear operator T from E to F is called a "compact operator" if T maps the unit ball of E into a compactoid.*

Prof. N. De Grande-De Kimpe suggested, during a personal discussion, that the concept analogous to nuclear spaces relevant to non-Archimedean analysis is that of a Schwartz space introduced below.

Definition 30. *A locally K-convex topological linear space E is said to be a Schwartz space, if given β there exists $\alpha > \beta$ such that the canonical mapping:*

$$\left(E \big/ \wp_\alpha^{-1}(0) \right)^{\sim} \to \left(E \big/ \wp_\beta^{-1}(0) \right)^{\sim}$$

obtained by extending the mapping:

$$x + \wp_\alpha^{-1}(0) \to x + \wp_\beta^{-1}(0), \ x \in E,$$

is a compact operator.

The following theorem characterizes Schwartz spaces among $c_0(p)$.

Theorem 52. *Let K be a non-trivially valued, non-Archimedean field, which is spherically complete. Then $c_0(p)$ is a Schwartz space if and only if*

$$\lim_{k \to \infty} p_k = 0.$$

The proof of the theorem depends on Theorem 53 below due to De Grande-De Kimpe's communication during a personal discussion. The statement of Theorem 53 requires the following concept of the Köthe-Toeplitz dual in non-Archimedean analysis, which is analogous to the concept of this dual consisting of complex sequences for a set P of real sequences. In the present case, it is convenient to take P to be a sequence $\{\alpha^{(i)}\}$ of real sequences $\{\alpha_k^{(i)}\}_{k=1}^{\infty}$, $i = 1, 2, \ldots$ satisfying the conditions:

(i) $\alpha_k^{(i)} > 0$, $i, k = 1, 2, \ldots$;

and

(ii) $\alpha_k^{(i+1)} \geq \alpha_k^{(i)}$, $i, k = 1, 2, \ldots$.

We note that such a P is a candidate for the P chosen in Section 5.3 to define the Köthe-Toeplitz dual in the classical case. Then the dual $\hat{P}$ is defined to be the set of sequences $\{x_k\}$, $x_k \in K$, $k = 1, 2, \ldots$ such that

$$\lim_{k \to \infty} |x_k| \alpha_k^{(i)} = 0, \; i = 1, 2, \ldots.$$

Then $\hat{P}$ is a complete, locally K-convex space with the non-Archimedean seminorms (in fact, norms)

$$\mathcal{P}_i(x) = \sup_{k \geq 1} |x_k| \alpha_k^{(i)}, \; x = \{x_k\} \in \hat{P}, i = 1, 2, \ldots.$$

With this background, we can now state and prove De Grande-De Kimpe's theorem. The proof is given for the sake of completeness.

Theorem 53 (De Grande-De Kimpe). *If K is spherically complete, then the following condition is necessary and sufficient for $\hat{P}$ to be a Schwartz space:*

$$\left. \begin{array}{c} \textit{For every } i = 1, 2, \ldots, \textit{ there exists } i' > i \textit{ such that} \\[2mm] \dfrac{\alpha_k^{(i)}}{\alpha_k^{(i')}} \to 0, \; k \to \infty. \end{array} \right\} \qquad (5.6)$$

Proof. Let $\hat{P}$ be a Schwartz space. Then for each i, there exists $i' > i$ such that the canonical mapping

$$\eta : (\hat{P}, \wp_{i'})^{\sim} \to (\hat{P}, \wp_i)^{\sim},$$

which is, in fact, the identity mapping is compact. Since K is non-trivially valued, choose $\pi \in K$ and integers $n(k)$ such that $0 < |\pi| < 1$ and

$$|\pi|^{n(k)+1} \leq \alpha_k^{(i')} < |\pi|^{n(k)}, \; k = 1, 2, \ldots.$$

Consider the sequences

$$\xi^{(j)} = \{0, 0, \ldots, 0, \frac{1}{\pi^{\xi(j)}}, 0, \ldots\},$$

the non-zero entry occurring in the jth place, $j = 1, 2, \ldots$. Clearly they are in the unit sphere U of $(\hat{P}, \wp_{i'})$. Hence $\overline{U}$, the closure of U in $(\hat{P}, \wp_i)^\sim$, is a compactoid. Thus, given $\epsilon > 0$, there exist a finite number of points $\mu^{(1)}, \ldots, \mu^{(n)}$ of $(\hat{P}, \wp_i)^\sim$ such that

$$\overline{U} \subset B_\epsilon(0) + \overline{co}\{\mu^{(1)}, \ldots, \mu^{(n)}\}.$$

Choose points $\mu_1^{(1)}, \ldots, \mu_1^{(n)}$ of $(\hat{P}, \wp_i)$ such that

$$\wp_i(\mu^{(j)} - \mu_1^{(j)}) < \epsilon, \ j = 1, 2, \ldots, n,$$

where $\wp_i$ also stands for its completion. It follows that

$$B_\epsilon(0) + \overline{co}\{\mu^{(1)}, \ldots, \mu^{(n)}\} = B_\epsilon(0) + \overline{co}\{\mu_1^{(1)}, \ldots, \mu_1^{(n)}\}.$$

Since $\mu_1^{(1)}, \ldots, \mu_1^{(n)}$ are elements of $\hat{P}$, there exists a positive integer k_0 such that for all $k \geq k_0$,

$$|\mu_{1k}^{(j)}| \alpha_k^{(i)} < \epsilon, \ j = 1, 2, \ldots, n$$

so that we can suppose that

$$\mu_1^{(j)} = \{\mu_{11}^{(j)}, \mu_{12}^{(j)}, \ldots, \mu_{1k_0}^{(j)}, 0, \ldots\}, \ j = 1, 2, \ldots.$$

Now, $\xi^{(j)}$ can be written as

$$\xi^{(j)} = \eta^{(j)} + \sum_{\ell=1}^{n} \lambda_{j\ell} \mu_1^{(\ell)}, \ \eta^{(j)} \in B_\epsilon(0), j = 1, 2, \ldots.$$

Thus, for $j > k_0$,

$$\frac{1}{|\pi|^{n(j)}} \alpha_j^{(i)} < \epsilon.$$

Consequently, for $j > k_0$,

$$\frac{\alpha_j^{(i)}}{\alpha_j^{(i')}} = \frac{\left(\frac{\alpha_j^{(i)}}{|\pi|^{n(j)}}\right)}{\left(\frac{\alpha_j^{(i')}}{|\pi|^{n(j)}}\right)}$$

$$\leq \frac{\epsilon}{|\pi|},$$

$$i.e., \ \frac{\alpha_k^{(i)}}{\alpha_k^{(i')}} \to 0, \ k \to \infty.$$

Conversely, suppose (5.6) holds. For any $\epsilon > 0$, there exists k_0 such that

$$\frac{\alpha_k^{(i)}}{\alpha_k^{(i')}} < \epsilon, \quad \text{for all } k > k_0.$$

We first prove that

$$\eta : (\hat{P}, \wp_{i'}) \to (\hat{P}, \wp_i)^{\sim},$$

which proves that the identity mapping is compact. To this end, it suffices to show that the image of the unit sphere U in $(\hat{P}, \wp_{i'})$ is a compactoid. On K^{k_0}, define $\| \ \|^*$ by

$$\|(\mu_1, \ldots, \mu_{k_0})\|^* = \max_{1 \le k \le k_0} |\mu_k| \alpha_k^{(i)}.$$

Clearly $\| \ \|^*$ is a norm on K^{k_0}. Since K^{k_0} is finite dimensional, using the fact that K is spherically complete, K^{k_0} is c-compact in the $\| \ \|^*$ topology. Hence the unit sphere is K^{k_0}, and being a closed subset, is also c-compact in the $\| \ \|^*$ topology. Thus the unit sphere is a compactoid ([19, p. 161, Theorem 4.56]). Hence there exist points

$$\mu^{(1)} = (\mu_1^{(1)}, \ldots, \mu_{k_0}^{(1)}), \ldots, \mu^{(n)} = (\mu_1^{(n)}, \ldots, \mu_{k_0}^{(n)})$$

such that the unit sphere in K^{k_0} is contained in

$$B_\epsilon'(0) + \overline{co}\{\mu^{(1)}, \ldots, \mu^{(n)}\},$$

where

$$B_\epsilon'(0) = \{x \in K^{k_0} : \|x\|^* \le \epsilon\}.$$

Now,

$$\nu^{(1)} = \{\mu_1^{(1)}, \ldots, \mu_{k_0}^{(1)}, 0, 0, \ldots\}, \ldots, \nu^{(n)} = \{\mu_1^{(n)}, \ldots, \mu_{k_0}^{(n)}, 0, 0, \ldots\}$$

are elements of $\hat{P}$. Let $\{\xi_j\}$ be any element of $\hat{P}$ such that $\wp_{i'}(\{\xi_j\}) \le 1$. Clearly

$$\{0, \ldots, 0, \xi_{k_0+1}, \xi_{k_0+2}, \ldots\} \in B_\epsilon(0) \text{ in } (\hat{P}, \wp_i)$$

and

$$\{\xi_1, \ldots, \xi_{k_0}, 0, 0, \ldots\} \in B_\epsilon(0) + \overline{co}\{\nu^{(1)}, \ldots, \nu^{(n)}\},$$

$$\text{i.e., } \{\xi_j\} \in B_\epsilon(0) + \overline{co}\{\nu^{(1)}, \ldots, \nu^{(n)}\}.$$

Thus we have proved that

$$\eta : (\hat{P}, \wp_{i'}) \to (\hat{P}, \wp_i)^\sim$$

is compact. It remains to prove that

$$\eta : (\hat{P}, \wp_{i'})^\sim \to (\hat{P}, \wp_i)^\sim$$

is compact. For, more generally, if X, Y are non-Archimedean normed linear spaces of which Y is complete and $f : X \to Y$ is compact, then f can be uniquely extended to a compact mapping $\tilde{f} : \tilde{X} \to Y$. To prove this, define

$$\tilde{f}(x) = \begin{cases} f(x), & \text{if } x \in X; \\ \lim_{n \to \infty} f(x_n), & \text{if } x \in \tilde{X} - X, x_n \to x. \end{cases}$$

If $\tilde{U}$ is the unit sphere in $\tilde{X}$, then $U = \tilde{U} \cap X$ is the unit sphere in X. By assumption, $f(U)$ is a compactoid. Thus there exist points $y_1, \ldots, y_n \in Y$ such that

$$f(U) \subset B_\epsilon(0) + \overline{co}\{y_1, \ldots, y_n\}.$$

If $x \in \tilde{U}$, $x_n \in X$, $x_n \to x$, $n \to \infty$, then $x_n \in \tilde{U}$, $n \geq n_0$ and so $x_n \in U$, $n \geq n_0$. Thus

$$\tilde{f}(x) = \lim_{n \to \infty} f(x_n) \in B_\epsilon(0) + \overline{co}\{y_1, y_2, \ldots, y_n\}.$$

Consequently $\tilde{f}(\tilde{U})$ is a compactoid and so $\tilde{f} : \tilde{X} \to Y$ is compact, completing the proof. □

Remark 16. *(i) Prof. De Grande-De Kimpe, in a personal communication, observed that the proof of Theorem 53 can be simplified using an her result ([3, p. 21]).*

(ii) The present remarks are intended to warn the reader about seemingly contradicting results in existing literature on nuclear and Schwartz spaces over valued fields. In the first instance, analogous to the same concept in the case

$K = \mathbb{R}$ *or* $\mathbb{C}$, *we can define* $f : E \to F$, *where* E, F *are non-Archimedean*

Banach spaces over the complete, non-trivially valued, non-Archimedean field

K, *to be nuclear if there exist* $y_n \in F$, $a_n \in E^*$, $n = 1, 2, \ldots$ *such that*

$\lim\limits_{n \to \infty} \|y_n\| \|a_n\| = 0$ *and* $f(x) = \sum\limits_{n=1}^{\infty} a_n(x) y_n$, $x \in E$. *Such a mapping is com-*

pact in the sense of Definition 29 (see e.g., [19, p. 143, Theorem 4.40]). An

immediate consequence of this observation is that the concepts of nuclear and

Schwartz spaces among locally K-*convex spaces are one and the same (cf. De*

Grande-De Kimpe's suggestion that the latter is relevant to non-Archimedean

analysis mentioned earlier), when K *is a complete, non-trivially valued, non-*

Archimedean field. van der Put and van Tiel ([18, p. 556, italicized statement])

claim that if K *is spherically complete, then any locally* K-*convex space is*

necessarily nuclear in the sense of Grothendieck ([4, p. 80, Definition 4]). It

then follows that such a space is nuclear with the definition of nuclear map-

pings suggested in the beginning of these remarks. Consequently it is also a

Schwartz space. It looks as if there is a contradiction to a result of De Grande-

De Kimpe ([2, p. 123]): there are not infinite dimensional non-Archimedean

Banach spaces which are Schwartz spaces. For, c_0, *with* K *spherically com-*

plete, should be a Schwartz space which is a Banach space, while it is not finite

dimensional. The seeming contradiction is because van der Put and van Tiel

[18] employed the analogue of Grothendieck's (loc. cit) definition of nuclear

spaces to define nuclear spaces over non-Archimedean valued fields. Because

of De Grande-De Kimpe's result (loc. cit), it turns out that what she calls a

Schwartz space is not the same as what van der Put and van Tiel [18] call

a nuclear space. The latter authors have given no indication of how their re-

sult could be established, while van Rooij ([19, p. 159]) gives a passing hint

for a possible deduction of this result. In this context, for further forthright

comments, the reader can refer to [1].

   We note that

$$c_0(p) = \hat{P},$$

where

$$P = \left\{ \{ i^{\frac{1}{p_k}} \}_{k=1}^{\infty}, i = 1, 2, \dots \right\}$$

as sets and locally K-convex spaces.

Proof of Theorem 52. Because of the observation made just above, Theorem 52 follows from Theorem 53, since

$$\frac{i^{\frac{1}{p_k}}}{i^{\frac{2}{p_k}}} = \frac{1}{i^{\frac{1}{p_k}}} \to 0, \; k \to \infty, i = 1, 2, \dots,$$

if $\lim\limits_{k \to \infty} p_k = 0$. Conversely, if $i' > i$, $i = 1, 2, \dots$ and $\left(\frac{i}{i'}\right)^{\frac{1}{p_k}} \to 0$, $k \to \infty$, it is obvious that $\lim\limits_{k \to \infty} p_k = 0$, completing the proof.

Remark 17. *If $P = \left\{ \left\{ \left(\frac{i}{i+1}\right)^{\frac{1}{p_k}} \right\}_{k=1}^{\infty}, i = 1, 2, \dots \right\}$ and K is spherically complete, then $\hat{P}$ is a Schwartz space if and only if*

$$\lim_{k \to \infty} p_k = 0.$$

Note that $\hat{P}$ can also be described as a $\Lambda_0(p)$, an analogue of a power series space of finite type defined by

$$\Lambda_0(p) = \left\{ \{x_k\} : |x_k| \left(\frac{i}{i+1}\right)^{\frac{1}{p_k}} \to 0, \; k \to \infty, i = 1, 2, \dots \right\}.$$

5.5 $c_0(p)$ as a metric linear algebra

Srinivasan [16] studied the ideal structure of $c_0(p)$, $p_k = \frac{1}{k}$, $k = 1, 2, \dots$ or what is the same as the space of entire functions, considered as an algebra without identity with the Hadamard product of sequences, viz., if $x = \{x_k\}$, $y = \{y_k\} \in c_0(p)$, $xy = \{x_k y_k\}$ when $K = \mathbb{R}$ or $\mathbb{C}$. Later he also studied [17] the analogous structure of $c_0(p)$, for the same choice of p, when K is a complete, non-trivially valued, non-Archimedean field. The object of the present section is to show that this study could be extended to $c_0(p)$ for any bounded, positive sequence $p = \{p_k\}$ and any complete, non-trivially valued field K. Throughout this section, we assume that K is such a field.

To begin with, we observe that $g(xy) \leq g(x)g(y)$, $x, y \in c_0(p)$ if g is the paranorm on $c_0(p)$ defined by (1.7). Consequently $c_0(p)$ is a metric linear algebra. In case K is a non-Archimedean field, the metric is also non-Archimedean. We define the binary operation $\oplus$ in $c_0(p)$ by

$$x \oplus y = x + y + xy, \ x, y \in c_0(p)$$

and recall (Definition 14) that $x \in c_0(p)$ is quasi-invertible when there exists $y \in c_0(p)$, called the quasi-inverse of x, such that $x \oplus y = 0$.

We observe that $x = \{x_k\} \in c_0(p)$ has a quasi-inverse $y = \{y_k\} \in c_0(p)$ if and only if $x_k \neq -1$ for all k. For, if $x \oplus y = 0$, i.e., $x_k + y_k + x_k y_k = 0$, $k = 1, 2, \ldots$, $x_k = -1$ for some k implies $-1 = 0$, which is absurd. Conversely, if $x_k \neq -1$, $k = 1, 2, \ldots$, $\{y_k\} \in c_0(p)$, if $y_k = -\frac{x_k}{x_k+1}$, $k = 1, 2, \ldots$ and $x \oplus y = 0$. Let

$$\mathscr{E} = \{e_1, e_2, \ldots, e_k, \ldots\}$$

be, as usual, the set of all unit sequences, where

$$e_k = \{0, \ldots, 0, 1, 0, \ldots\},$$

1 occurring in the kth place, $k = 1, 2, \ldots$. It may be noted that $\mathscr{E}$ is a Schauder basis for $c_0(p)$. If $I \subset c_0(p)$ is an ideal, let $\mathscr{I}$ denote the set of all $e_i \in \mathscr{E}$ such that $e_i \in I$. Thus every ideal I of $c_0(p)$, which is not the singleton $\{0\}$, determines a subset $\mathscr{I}$ of $\mathscr{E}$. Non-zero closed ideals of $c_0(p)$ are related to the subsets of $\mathscr{E}$ determined by them by the following theorem.

Theorem 54. *(i) Every non-zero subset I of $c_0(p)$ is a closed ideal of $c_0(p)$ if and only if it is the closed linear span of $\mathscr{I}$.*

(ii) If, in addition, the closed ideal is to be maximal, it should be precisely the closed linear span of $\mathscr{E} - \{e_i\}$ for some i.

Proof. (i) If I is a non-zero closed ideal of $c_0(p)$ and $x = \{x_k\} \in I$, $\{x_k\} = \sum_{e_k \in \mathscr{I}} x_k e_k$. Thus $\{x_k\} \in \overline{[\mathscr{I}]}$, the closed linear span of $\mathscr{I}$. Now, $\mathscr{I} \subset I$ implies $\overline{[\mathscr{I}]} \subset I$ since I is closed. Thus $I = \overline{[\mathscr{I}]}$, if I is a closed ideal. Conversely, $\overline{[\mathscr{I}]}$ is a closed ideal for any subset $\mathscr{I}$ of $\mathscr{E}$. For, if $x = \{x_k\} \in c_0(p)$, $y = \{y_k\} \in \overline{[\mathscr{I}]}$ and $\{x_k\} = \sum_{i=1}^{\infty} x_i e_i$, $\{y_k\} = \sum_{e_i \in \mathscr{I}} y_i e_i$, then $xy = \sum_{e_i \in \mathscr{I}} x_i y_i e_i \in \overline{[\mathscr{I}]}$. $\overline{[\mathscr{I}]}$ is a closed subspace already and hence a closed ideal of $c_0(p)$.

(ii) Let I be a maximal closed ideal of $c_0(p)$ and $I \subsetneq c_0(p)$. If possible, let there be two elements of $\mathscr{E}$, (say) $e_i, e_j \notin \mathscr{I}$. Then $I \subset \overline{[\mathscr{I} \cup \{e_i\}]} \neq c_0(p)$, which contradicts the fact that I is maximal. Thus

$$I = \overline{[\mathscr{E} - \{e_i\}]} \quad \text{for some } i.$$

Conversely, if $\overline{[\mathscr{I}]} = \overline{[\mathscr{E} - \{e_i\}]} \subset I'$, where I' is a closed ideal of $c_0(p)$ and $x = \{x_k\} \in I' - \overline{[\mathscr{I}]}$, $\{x_k\} = \sum_{e_k \in \mathscr{I}'} x_k e_k$, then $x_i \neq 0$. For, otherwise $x \in \overline{[\mathscr{I}]}$. Also $xe_i \in I'$, I' being an ideal; so $e_i = x_i^{-1} xe_i \in I'$. Thus $\mathscr{I}' = \mathscr{E}$ and so $I' = c_0(p)$, i.e., $\overline{[\mathscr{I}]}$ is a maximal closed ideal, completing the proof. $\qquad\square$

Corollary 4. $c_0(p)$ *is semi-simple.*

We recall (Definition 15) that an ideal I of $c_0(p)$ is regular if there exists $u \in c_0(p)$ such that $ux - x \in I$ for every $x \in c_0(p)$.

The following result connects closed ideals with regular ideals.

Theorem 55. *An ideal of $c_0(p)$ is a maximal regular ideal if and only if it is a maximal closed ideal.*

Proof. Let $I \subsetneq c_0(p)$ be a maximal regular ideal of $c_0(p)$. Then there exists $u \in c_0(p)$ such that $ux - x \in I$ for every $x \in c_0(p)$. We first observe that for no $x \in I$ is $g(u - x) < 1$. For, if $g(u - x) < 1$ for some $x \in I$, $x - u$ is quasi-invertible in the sense of Definition 14. If $v \in c_0(p)$ is the quasi-inverse of $x - u$, $v + x - u + v(x - u) = 0$. Since $v - vu \in I$ and $x + vx \in I$, I being a regular ideal, $u \in I$ and so $I = c_0(p)$ and is not proper, a contradiction. Now, $u \notin \bar{I}$. Otherwise, there exists $x \in I$ such that $g(x - u) < 1$, contradicting what has been established just now. Thus

$$I \subset \overline{I} \subsetneq c_0(p).$$

Since I is maximal and $\overline{I}$ is a regular ideal, $\overline{I} = I$ and so I is closed. Since I is also regular, it follows that I is a maximal closed ideal. Conversely, if $I \subsetneq c_0(p)$ is a maximal closed ideal of $c_0(p)$ and if $I = \overline{[\mathscr{I}]}$, where $\mathscr{I} = \mathscr{E} - \{e_i\}$ for some i, $e_i x - x \in I$ for every $x \in c_0(p)$ and so I is a regular ideal. I is a maximal regular ideal or is contained in one such ideal. The latter possibility contradicts the maximal closed nature of I by what has been proved earlier, completing the proof. $\square$

Theorem 56. *In $c_0(p)$, every closed ideal I is the intersection of all maximal closed ideals containing I.*

Proof. Let I be any closed ideal of $c_0(p)$. Then $I = \overline{[\mathscr{I}]}$. Let

$$\Lambda = \{k : M = \overline{[\mathscr{E} - \{e_k\}]}, \ M \supset I\}.$$

It suffices to show that $\bigcap_{k \in \Lambda} \overline{[\mathscr{E} - \{e_k\}]} \subset I$. To this end, we observe that $\bigcap_{k \in \Lambda} [\mathscr{E} - \{e_k\}] \subset \mathscr{I}$. For, if $e_{k_0} \in \bigcap_{k \in \Lambda} [\mathscr{E} - \{e_k\}]$, $e_{k_0} \notin \mathscr{I}$, then $e_{k_0} \notin I$, $I \subset \overline{[\mathscr{E} - \{e_{k_0}\}]}$ and consequently, $e_{k_0} \in \mathscr{E} - \{e_{k_0}\}$, which is a contradiction. Now,

$$\bigcap_{k \in \Lambda} [\mathscr{E} - \{e_k\}] = \mathscr{E} - \{e_k : k \in \Lambda\}.$$

Thus, if $x \in \bigcap_{k \in \Lambda} \overline{[\mathscr{E} - \{e_k\}]}$, $x = \sum_{i=1}^{\infty} x_i e_i$, where $x_i = 0$ for every $i \in \Lambda$. Hence

$$x \in \overline{[\mathscr{E} - \{e_k : k \in \Lambda\}]}$$

$$= \overline{\left[\bigcap_{k \in \Lambda} \{\mathscr{E} - \{e_k\}\} \right]}$$

$$\subset \overline{[\mathscr{I}]}$$

$$= I,$$

completing the proof. $\square$

5.6 Step spaces

An alternative characterization of closed ideals in $c_0(p)$ can be given in terms of step spaces introduced in Definition 11.

We observed that each maximal closed ideal of $c_0(p)$ is a step space and every step space of $c_0(p)$ is a closed ideal. If $\mathcal{M}$ denotes the set of all maximal closed ideals of $c_0(p)$ containing a given closed ideal I and $\mathcal{S}$ denotes the set of all step spaces containing I, then

$$\bigcap_{S \in \mathcal{S}} S \subset \bigcap_{M \in \mathcal{M}} M = I \subset \bigcap_{S \in \mathcal{S}} S,$$

from which it follows that

$$I = \bigcap_{S \in \mathcal{S}} S.$$

Thus we have the following result.

Theorem 57. *Every closed ideal of $c_0(p)$ is the intersection of all step spaces containing it.*

5.7 Some more properties of the sequence space $c_0(p)$

In this section, we prove some more results about the space $c_0(p)$ over a complete, non-trivially valued, non-Archimedean field K, which extend some of the results for the space of entire functions proved by Raghunathan [13, 14]. For the details recorded in this section, one can refer to Natarajan [11]. We write sequences as $\{x_n\}$, $n = 1, 2, \ldots$, for convenience.

Theorem 58. *If $\lim_{n \to \infty} p_n = 0$, $c_0(p)$ is the unique maximal linear metric subspace of $c_0^*(p) = \ell_\infty(p)$.*

Proof. Note that $c_0(p)$ is a linear metric subspace of $c_0^*(p) = \ell_\infty(p)$. Let, now, S be a linear metric subspace of $c_0^*(p)$. We claim that $S \subseteq c_0(p)$. If not, there exists $c = \{c_n\} \in S$, $c = \{c_n\} \notin c_0(p)$. So

$$|c_n|^{p_n} \not\to 0, \ n \to \infty.$$

By hypothesis, $\lim_{n\to\infty} p_n = 0$. Consequently, there exists a strictly increasing sequence $\{n(q)\}$ of positive integers such that

$$|c_{n(q)}|^{p_{n(q)}} > \epsilon$$

and

$$p_{n(q)} < \frac{1}{q}, \ q = 1, 2, \ldots.$$

Since K is non-trivially valued, there exists $\pi \in K$ such that $0 < \rho = |\pi| < 1$. Let the sequence $\{a_n\}$ be defined by

$$a_n = \left.\begin{matrix} \pi^q, & \text{if } n = n(q); \\ \\ 0, & \text{if } n \neq n(q) \end{matrix}\right\}, \ q = 1, 2, \ldots.$$

It is clear that $\lim_{n\to\infty} a_n = 0$. However, for $q = 1, 2, \ldots$,

$$g(a_{n(q)}c) = \sup_{n\geq 1} |a_{n(q)}c_n|^{p_n}$$

$$\geq |a_{n(q)}c_{n(q)}|^{p_{n(q)}}$$

$$> \epsilon|a_{n(q)}|^{p_{n(q)}}$$

$$= \epsilon\rho^{qp_{n(q)}}$$

$$> \epsilon\rho,$$

since $qp_{n(q)} < 1$ and $0 < q < 1$ and so $\rho^{qp_{n(q)}} > \rho$. Thus $a_n c \not\to 0$, $n \to \infty$, though $a_n \to 0$, $n \to \infty$. Consequently, S is not a linear metric subspace of $c_0^*(p)$, which is a contradiction, completing the proof of the theorem. $\square$

Let $a = \{a_n\} \in c_0(p)$. For any $r \in |K^+|$, the value group of K, define

$$\|a\|_r = \sup_{n\geq 1}\{|a_n|r^{\frac{1}{p_n}}\}. \tag{5.7}$$

We note that $\| \ \|_r$ is a non-Archimedean norm on $c_0(p)$. We shall denote the non-Archimedean normed linear space $(c_0(p), \| \cdot \|_r)$ by $(c_0(p))_r$. If $r_1 > r_2$, $\|a\|_{r_1} > \|a\|_{r_2}$ and so $(c_0(p))_{r_1}$ is weaker than $(c_0(p))_{r_2}$ as topological spaces. We also note that for any $r \in |K^+|$, $(c_0(p))_r$ is stronger than $c_0(p)$. Thus, as r increases, $(c_0(p))_r$ form a decreasing family of non-Archimedean normed linear spaces, each of which is stronger than $c_0(p)$.

We need the following lemma.

Lemma 4. *If $g(a) \geq d > 0$, then*

$$\|a\|_r \geq d$$

for any $r \in |K^+|$ such that $r > \frac{D}{d}$, where $D = \max(1, d^H)$, $H = \sup\limits_{n \geq 1} p_n$.

Proof. Let $a = \{a_n\} \in c_0(p)$ with $g(a) \geq d > 0$. If possible, let $\|a\|_r < d$ for some $r \in |K^+|$. Then $\sup\limits_{n \geq 1} |a_n| r^{\frac{1}{p_n}} < d$ and so

$$|a_n|^{p_n} r < d^{p_n} < D.$$

If $D < dr$, then

$$|a_n|^{p_n} < \frac{D}{r} < d, \ n = 1, 2, \ldots.$$

Thus

$$g(a) < d,$$

which is a contradiction. So

$$D \geq dr$$

$$i.e., \ r \leq \frac{D}{d}.$$

Consequently, if $r > \frac{D}{d}$, $\|a\|_r \geq d$, completing the proof. $\qquad\square$

As an immediate consequence of Lemma 4, we have the following result.

Theorem 59. *Let $S \subseteq c_0(p)$. Then*

$$(\overline{S})_{c_0(p)} = \bigcap_{r \in |K^+|} (\overline{S})_{(c_0(p))_r},$$

where $(\overline{S})_{c_0(p)}$ is the closure of S in the space $c_0(p)$, etc.

Proof. Since $(c_0(p))_r$ is stronger than $c_0(p)$, $r \in |K^+|$,

$$(\overline{S})_{c_0(p)} \subseteq (\overline{S})_{(c_0(p))_r}, \ r \in |K^+|.$$

Consequently,

$$(\overline{S})_{c_0(p)} \subseteq \bigcap_{r \in |K^+|} (\overline{S})_{(c_0(p))_r}.$$

The reverse inclusion follows from Lemma 4, completing the proof. $\square$

Theorem 60. *Every functional in* $(c_0(p))_r^*$ *is given by*

$$f(a) = \sum_{n=1}^{\infty} a_n c_n, \ a = \{a_n\} \in c_0(p), \text{ where } \left\{ \frac{|c_n|}{r^{\frac{1}{p_n}}} \right\} \tag{5.8}$$

is bounded and conversely.

Proof. Let $f \in (c_0(p))_r^*$. Then

$$|f(a)| \leq M \|a\|_r,$$

$M > 0$, $a = \{a_n\} \in c_0(p)$. Let $e_n = \{0, \dots, 0, 1, 0, \dots\}$, 1 occurring in the nth place, 0's elsewhere, $n = 1, 2, \dots$. Let $0 < \epsilon < 1$. Since $a \in c_0(p)$, there exists a positive integer N such that

$$|a_n|^{p_n} < \frac{\epsilon^H}{r}, \ n > N, H = \sup_{n \geq 1} p_n,$$

$$i.e., \ |a_n| < \left(\frac{\epsilon^H}{r} \right)^{\frac{1}{p_n}}, \ n > N.$$

So, if $n > N$,

$$\left\| a - \sum_{k=1}^{n} a_k e_k \right\|_r = \|\{0, \dots, 0, a_{n+1}, a_{n+2}, \dots\}\|_r$$

$$= \sup_{k \geq n+1} |a_k| r^{\frac{1}{p_k}}$$

$$< \sup_{k \geq n+1} \epsilon^{\frac{H}{p_k}}$$

$$< \epsilon, \quad \text{since } 0 < \epsilon < 1 \text{ and } \frac{H}{p_k} \geq 1.$$

Hence,

$$a = \sum_{n=1}^{\infty} a_n e_n$$

and so

$$f(a) = \sum_{n=1}^{\infty} a_n f(e_n).$$

Choosing $a = e_n$, we have,

$$|f(e_n)| \le M \|e_n\|_r = M r^{\frac{1}{p_n}}, \ n = 1, 2, \ldots.$$

Taking $c_n = f(e_n)$, we have that $\left\{ \frac{|c_n|}{r^{\frac{1}{p_n}}} \right\}$ is bounded. Conversely, let (5.8) hold. Let $M > 1$ be such that

$$\frac{|c_n|}{r^{\frac{1}{p_n}}} \le M, \ n = 1, 2, \ldots.$$

Now,

$$|a_n c_n|^{p_n} \le |a_n|^{p_n} M^{p_n} r$$

$$\le |a_n|^{p_n} M^H r$$

$$\to 0, \ n \to \infty, \ \text{since } \{a_n\} \in c_0(p).$$

Since $c_0(p) \subseteq c_0$, $a_n c_n \to 0$, $n \to \infty$ and so $\sum_{n=1}^{\infty} a_n c_n$ converges. Thus

$$f(a) = \sum_{n=1}^{\infty} a_n c_n$$

is defined. It is clear that f is linear. Also,

$$|f(a)| \le \sup_{n \ge 1} |a_n c_n|$$

$$\le M \sup_{n \ge 1} |a_n| r^{\frac{1}{p_n}}$$

$$= M \|a\|_r,$$

which shows that f is bounded. Thus $f \in (c_0(p))_r^*$, completing the proof. $\quad\square$

Theorem 61. $c_0^*(p) = \bigcup_{r \in |K^+|} (c_0(p))_r^*.$

Proof. Since, $(c_0(p))_r$ is stronger than $c_0(p)$, for every $r \in |K^+|$,

$$\bigcup_{r \in |K^+|} (c_0(p))_r^* \subseteq c_0^*(p).$$

To prove the reverse inclusion, let $f \in c_0^*(p)$. Then

$$f(a) = \sum_{n=1}^{\infty} a_n c_n, \ a = \{a_n\} \in c_0(p), \{c_n\} \in \ell_\infty(p).$$

So

$$|c_n|^{p_n} \leq M, \ n = 1, 2, \dots, M > 0.$$

Since K is non-trivially valued, there exists $r \in |K^+|$ such that $0 < M \leq r$. Hence $|c_n|^{p_n} \leq r, \ n = 1, 2, \dots$. Thus, $\left\{ \dfrac{|c_n|}{r^{\frac{1}{p_n}}} \right\}_{n=1}^{\infty}$ is bounded so that $f \in (c_0(p))_r^*$ and so

$$c_0^*(p) \subseteq \bigcup_{r \in |K^+|} (c_0(p))_r^*,$$

completing the proof of the theorem. $\square$

Remark 18. *Theorem 61 is equivalent to the following result, which we recall in this context:*

The continuous dual $c_0^*(p)$ of $c_0(p)$ consists of all functionals f given by:

$$f(a) = \sum_{k=1}^{\infty} a_k c_k, \ a = \{a_k\} \in c_0(p), c = \{c_k\} \in \ell_\infty(p).$$

In other words, $c_0^*(p) = \ell_\infty(p)$, upto isometric isomorphism.

We conclude by remarking that the results proved in this section hold good when $K = \mathbb{R}$ or $\mathbb{C}$ also. We leave the details to the reader.

Bibliography

[1] R. Bhaskaran and P.N. Natarajan. The space $c_0(p)$ over valued fields. *Rocky Mountain J. Math.*, 16:129–136, 1986.

[2] N. De Grande-De Kimpe. On spaces of operators between locally K-convex spaces. *Indag. Math.*, 34:113–129, 1972.

[3] N. De Grande-De Kimpe. Structure theorems for locally K-convex spaces. *Indag. Math.*, 39:11–22, 1977.

[4] A. Grothendieck. *Produits tensoriels topologiques et espaces nucléaires*. *Memoirs A.M.S.*, 16, 1955.

[5] G. Köthe. *Topological vector spaces I*. Springer, 1969.

[6] C.G. Lascarides. A study of certain sequence spaces of Maddox and a generalization of a theorem of Iyer. *Pacific J. Math.*, 38:487–500, 1971.

[7] I.J. Maddox. Continuous and Köthe-Toeplitz duals of certain sequence spaces. *Proc. Cambridge Philos. Soc.*, 65: 431–435, 1969.

[8] I.J. Maddox and J.W. Roles. Absolute convexity in spaces of strongly summable sequences. *Canad. Math. Bull.*, 18:67–75, 1975.

[9] A.F. Monna. Espaces Vectoriels topologiques sur un corps valué. *Indag. Math.*, 24:351–367, 1962.

[10] P.N. Natarajan. The Steinhaus theorem for Toeplitz matrices in non-archimedean fields. *Comment. Math. Prace Mat.*, 20:417–422, 1978.

[11] P.N. Natarajan. More properties of $c_0(p)$ over non-archimedean fields. *J. Combinatorics Information System Sci.*, 38:121–127, 2013.

[12] A. Pietsch. *Nuclear locally convex spaces*. Springer, 1972.

[13] T.T. Raghunathan. On the space of entire functions over certain non-archimedean fields. *Boll. Un. Mat. Ital.*, 4:517–526, 1968.

[14] T.T. Raghunathan. On the space of entire functions over certain non-archimedean fields and its dual. *Studia Math.*, 33:251–256, 1969.

[15] H.H. Schaefer. *Topological vector spaces*. Springer, 1971.

[16] V.K. Srinivasan. On the ideal structure of the algebra of integral functions. *Proc. Nat. Inst. Sci. India, Part A*, 31:368–374, 1965.

[17] V.K. Srinivasan. On the ideal structure of the algebra of entire functions over complete, non-archimedean valued fields. *Arch. Math.*, 24:505–512, 1973.

[18] M. van der Put and J. van Tiel. Espaces nucléaires non-archimédiens. *Indag. Math.*, 29:556–561, 1967.

[19] A.C.M. Van Rooij. *Non-archimedean functional analysis*. Marcel Dekker, 1978.

[20] J. Van Tiel. Espaces localement K-convexes I-III. *Indag Math.*, 27:249–258; 259–272; 273–289, 1965.

Chapter 6

On the Sequence Spaces $\ell(p)$, $c_0(p)$, $c(p)$, $\ell_\infty(p)$ over Non-Archimedean Fields

6.1 Introduction

The sequence spaces $\ell(p)$, $c_0(p)$, $c(p)$ and $\ell_\infty(p)$ were introduced in Chapter 1. Simons [12], Maddox [3], and Bhaskaran and Natarajan [1] studied these spaces when the entries of the sequences are complex and $p = \{p_k\}$ is a sequence of positive real numbers such that $0 < \sup_{k\geq 1} p_k < \infty$. In this chapter, we record briefly some facts relating to continuous duals of these spaces and matrix transformations between these spaces, when the sequences have entries in a complete, non-trivially valued, non-Archimedean field K. It is to be noted that these spaces are linear spaces if and only if $\sup_{k\geq 1} p_k < \infty$. Further, since K is non-trivially valued,

$$c_0(p) \subset c_0, c(p) \subset c, \ell_\infty \subset \ell_\infty(p).$$

We also record some more properties of the above sequence spaces in this chapter. We also study the algebra $(\ell_\alpha, \ell_\alpha)$, $\alpha \geq 1$, in the context of a convolution product and prove a Mercerian theorem supplementing Theorem 29.

Throughout the present chapter, K is assumed to be a complete, non-trivially valued, non-Archimedean field.

6.2 Continuous duals and the related matrix transformations

For the contents of this section, the reader can refer to [6].

We recall that, if E is a sequence space, its generalized Köthe-Toeplitz dual, denoted by $E^\times$, is defined as

$$E^\times = \{x = \{x_k\} : x_k y_k \to 0, k \to \infty \text{ for all } y = \{y_k\} \in E\}.$$

It is easily proved that

$$c_0^\times(p) = \ell^\times(p) = \ell_\infty(p);$$
$$\ell_\infty^\times(p) = c_0(p);$$
$$and$$
$$c^\times(p) = c_0.$$

The continuous dual of E is denoted by E^*. Let X be a paranormed space with paranorm g (cf. Section 1.6). We note that a linear functional A on X is in X^* if and only if

$$\|A\|_M = \sup\{|A(x)| : g(x) \leq \frac{1}{M}\}$$

is finite for some $M > 1$.

We also note that $\ell(p), c_0(p), c(p)$ are linear topological spaces, while $\ell_\infty(p)$ is so if and only if

$$\inf_{k \geq 1} p_k > 0,$$

where the paranorms giving the respective topologies are:

$$\ell(p) : g(x) = \left(\sum_{k=1}^{\infty} |x_k|^{p_k}\right)^{\frac{1}{H}};$$

$$c_0(p), c(p), \ell_\infty(p) : g(x) = \sup_{k \geq 1} |x_k|^{\frac{p_k}{H}},$$

$H = \max(1, \sup_{k\geq 1} p_k)$.

The following results (see [6]) are easily proved (see Maddox [3] for the classical analogues).

Theorem 62. $f \in \ell^*(p)$ *if and only if*

$$f(x) = \sum_{k=1}^{\infty} a_k x_k, \{a_k\} \in \ell_\infty(p) \quad \text{for all} \quad x = \{x_k\} \in \ell(p).$$

Further, $\ell^*(p)$ is linearly homeomorphic to $\ell_\infty(p)$.

Theorem 63. $f \in c_0^*(p)$ *if and only if*

$$f(x) = \sum_{k=1}^{\infty} a_k x_k, \{a_k\} \in \ell_\infty(p) \quad \text{for all} \quad x = \{x_k\} \in c_0(p).$$

Further, $c_0^*(p)$ and ℓ_∞ are isometrically isomorphic if $\inf_{k\geq 1} p_k > 0$.

Theorem 64. $f \in c^*(p)$ *if and only if*

$$f(x) = \sum_{k=1}^{\infty} a_k x_k, \{a_k\} \in \ell_\infty \quad \text{for all} \quad x = \{x_k\} \in c(p).$$

Moreover, $c^*(p)$ and ℓ_∞ are isometrically isomorphic if $\inf_{k\geq 1} p_k > 0$.

The following lemmas are proved like their analogues for complex paranormed spaces (see [5]).

Lemma 5. *Let X be a paranormed space over K. Let $A_n \in X^*$, $n = 1, 2, \ldots$ and $q = \{q_n\}$ be bounded, $q_n > 0$, $n = 1, 2, \ldots$. Then*

$$\sup_{n\geq 1} \|A_n\|_M^{q_n} < \infty \quad \text{for some} \quad M > 1 \tag{6.1}$$

implies that

$$\{A_n(x)\} \in \ell_\infty(q), \ x \in X. \tag{6.2}$$

Further, if X is complete, (6.2) implies (6.1).

Lemma 6. *Let X be a paranormed space over K. Let $A_n \in X^*$, $n = 1, 2, \ldots$. If X has a fundamental set G, i.e., X is the closure of the linear span of G and $q = \{q_n\}$ is bounded, $q_n > 0$, $n = 1, 2, \ldots$, then*

$$\{A_n(b)\} \in c_0(q), \ b \in G \tag{6.3}$$

and

$$\lim_{M \to \infty} \limsup_{n \to \infty} (\|A_n\|_M)^{q_n} = 0 \tag{6.4}$$

together imply

$$\{A_n(x)\} \in c_0(q), \ x \in X. \tag{6.5}$$

In particular, if $q_n \to 0$, $n \to \infty$, (6.4) implies (6.5). If X is complete, (6.5) implies (6.4), even if q is unbounded.

We now prove the following theorems on matrix transformations between the spaces $\ell(p), c_0(p), c(p)$ and $\ell_\infty(p)$.

Theorem 65. $A = (a_{nk}) \in (c_0(p), c_0(q))$ *if and only if*

$$|a_{nk}|^{q_n} \to 0, \ n \to \infty, k = 1, 2, \ldots; \tag{6.6}$$

and

$$\left. \begin{array}{l} (i) \ \sup_{n \geq 1} \left(\sup_{k \geq 1} |a_{nk}| B^{-\frac{1}{p_k}} \right) < \infty \ \textit{for some } B > 1, \\[12pt] (ii) \ \lim_{M \to \infty} \limsup_{n \to \infty} \left(\sup_{k \geq 1} |a_{nk}| M^{-\frac{1}{p_k}} \right)^{q_n} = 0. \end{array} \right\} \tag{6.7}$$

Proof. If $A \in (c_0(p), c_0(q))$, $\{a_{nk}\}_{k=1}^{\infty} \in \ell_\infty(p)$, $n = 1, 2, \ldots$ (see [10]), so that (6.7) (i) is necessary. Also, $A_n \in c_0^*(p)$, $n = 1, 2, \ldots$ and $c_0(p)$ is complete. Hence by Lemma 6,

$$\lim_{M \to \infty} \limsup_{n \to \infty} (\|A_n\|_M)^{q_n} = 0.$$

So, for any $\epsilon > 0$, there exists $M > 1$ and a positive integer n_0 such that

$$\|A_n\|_M^{q_n} < \epsilon, \ n \geq n_0,$$

i.e., $|A_n(x)|^{q_n} < \epsilon$, $n \geq n_0$, for every $x \in c_0(p)$ such that $g(x) \leq \dfrac{1}{M}$, where g is the paranorm on $c_0(p)$. Now, for every $n = 1, 2, \ldots$, there exists an integer α_n such that

$$\rho^{\alpha_n + 1} \leq \epsilon^{\frac{1}{q_n}} < \rho^{\alpha_n},$$

where $\pi \in K$ is such that $0 < \rho = |\pi| < 1$, noting that such an element π exists since K is non-trivially valued.

If $n \geq n_0$,

$$|\pi^{-\alpha_n} A_n(x)| = |A_n(x)| \rho^{-\alpha_n}$$

$$< |A_n(x)| \epsilon^{-\frac{1}{q_n}}$$

$$< 1,$$

if $g(x) \leq \frac{1}{M}$. For $x = \{x_k\} \in c_0(p)$, let $\lambda \in K$ with $|\lambda| \geq 1$ such that $g(\lambda^{-1} x) \leq \frac{1}{M}$. Then

$$\left| \sum_{k=1}^\infty \pi^{-\alpha_n} a_{nk} \lambda^{-1} x_k \right| < 1.$$

Consequently, for $n \geq n_0$,

$$\left| \sum_{k=1}^\infty \pi^{-\alpha_n} a_{nk} x_k \right| < |\lambda|.$$

If $B = (b_{nk})$, $b_{nk} = \pi^{-\alpha_n} a_{nk}$, $n, k = 1, 2, \ldots$,

$$Bx = \left\{ \sum_{k=1}^\infty b_{nk} x_k \right\}_{n=1}^\infty \in \ell_\infty,$$

for every $x = \{x_k\} \in c_0(p)$ so that (see [10])

$$\sup_{n,k} (\rho^{-\alpha_n} |a_{nk}|)^{p_k} < \infty.$$

Hence

$$\sup_{n,k} (\epsilon^{-\frac{1}{q_n}} \rho |a_{nk}|)^{p_k} < L, \text{ for some } L > 1.$$

Then

$$\sup_{k \geq 1} (|a_{nk}| L^{-\frac{1}{p_k}}) < \rho^{-1} \epsilon^{\frac{1}{q_n}}, \ n = 1, 2, \ldots,$$

$$i.e., \ \left(\sup_{k \geq 1} |a_{nk}| L^{-\frac{1}{p_k}} \right)^{q_n} < \rho^{-q_n} \epsilon$$

$$< \rho^{-U} \epsilon, \ n = 1, 2, \ldots,$$

where $U = \sup\limits_{n \geq 1} q_n$. So condition (6.7) (ii) is necessary. It is clear that (6.6) is also necessary.

To prove the sufficiency of the conditions, we note that $\{e_k\}$, where e_k is the sequence with 1 at the kth place and 0 elsewhere, is a basis of $c_0(p)$ which is complete. By condition (6.7) (i),

$$\{a_{nk}\}_{k=1}^{\infty} \in \ell_{\infty}(p) = c_0^{\times}(p), \ n = 1, 2, \ldots.$$

Thus $(Ax)_n = \sum\limits_{k=1}^{\infty} a_{nk} x_k$ is defined for every $n = 1, 2, \ldots$. By (6.6),

$$\{A_n(e_k)\} \in c_0(q), \ k = 1, 2, \ldots.$$

By (6.7) (ii), given $\epsilon > 0$, there exist $M_0 > 1$ and a positive integer N such that

$$\left(\sup_{k \geq 1} |a_{nk}| M^{-\frac{1}{p_k}} \right)^{q_n} < \epsilon, \ n \geq N, M > M_0. \tag{6.8}$$

If now, $x = \{x_k\} \in c_0(p)$ is such that $g(x) \leq \frac{1}{M}$, it follows that

$$M^{\frac{1}{p_k}} |x_k| \leq 1, \ k = 1, 2, \ldots.$$

Hence

$$|A_n(x)| \leq \sup_{k \geq 1} |a_{nk}||x_k|$$

$$= \sup_{k \geq 1} |a_{nk}| M^{-\frac{1}{p_k}} M^{\frac{1}{p_k}} |x_k|$$

$$\leq \sup_{k \geq 1} |a_{nk}| M^{-\frac{1}{p_k}}.$$

In view of (6.8), for $M > M_0$, we have,

$$|A_n(x)|^{q_n} \leq \left(\sup_{k \geq 1} |a_{nk}| M^{-\frac{1}{p_k}} \right)^{q_n}$$

$$< \epsilon, \ n \geq N,$$

$x = \{x_k\} \in c_0(p)$ with $g(x) \leq \frac{1}{M}$. In other words,

$$\lim_{M \to \infty} \limsup_{n \to \infty} (\|A_n\|_M)^{q_n} = 0.$$

Appealing to Lemma 6, we conclude that $\{A_n(x)\} \in c_0(q)$ for every $x = \{x_k\} \in c_0(p)$, i.e., $A \in (c_0(p), c_0(q))$. This completes the proof of the theorem.

$\square$

In a similar manner, we can establish the following results using Lemma 5 and Lemma 6.

Theorem 66. $A = (a_{nk}) \in (\ell(p), \ell_\infty(q))$ *if and only if*

$$\sup_{n \geq 1} \left(\sup_{k \geq 1} |a_{nk}| M^{-\frac{1}{p_k}} \right)^{q_n} < \infty \text{ for some } M > 1. \tag{6.9}$$

Theorem 67. $A = (a_{nk}) \in (\ell(p), c_0(q))$ *if and only if (6.6) and (6.7) hold.*

Theorem 68. $A = (a_{nk}) \in (\ell(p), c(q))$ *if and only if (6.7) (i) holds; there exist $\alpha_k \in K$, $k = 1, 2, \ldots$ such that*

$$|a_{nk} - \alpha_k|^{q_n} \to 0, \ n \to \infty, k = 1, 2, \ldots \tag{6.10}$$

and

$$\lim_{M \to \infty} \limsup_{n \to \infty} \left(\sup_{k \geq 1} |a_{nk} - \alpha_k| M^{-\frac{1}{p_k}} \right)^{q_n} = 0. \tag{6.11}$$

Theorem 69. $A = (a_{nk}) \in (c_0(p), \ell_\infty(q))$ *if and only if (6.9) holds.*

Theorem 70. $A = (a_{nk}) \in (c_0(p), c(q))$ *if and only if (6.7) (i), (6.10) and (6.11) hold.*

Theorem 71. $A = (a_{nk}) \in (c(p), \ell_\infty(q))$ *if and only if*

$$a_{nk} \to 0, \ k \to \infty, n = 1, 2, \ldots \tag{6.12}$$

and (6.9) holds.

Theorem 72. $A = (a_{nk}) \in (c(p), c_0(q))$ *if and only if (6.6), (6.7) hold and*

$$\left| \sum_{k=1}^{\infty} a_{nk} \right|^{q_n} \to 0, \ n \to \infty. \tag{6.13}$$

Theorem 73. $A = (a_{nk}) \in (c(p), c(q))$ *if and only if (6.7) (i), (6.10), and (6.11) hold and there exists $\alpha \in K$ such that*

$$\left| \sum_{k=1}^{\infty} a_{nk} - \alpha \right|^{q_n} \to 0, \ n \to \infty. \tag{6.14}$$

The reader can try other matrix transformations involving the sequence spaces $\ell(p), c_0(p), c(p)$ and $\ell_\infty(p)$.

6.3   Some more properties of the sequence spaces $\ell(p), c_0(p), c(p)$ and $\ell_\infty(p)$

A sequence space E is said to be Köthe-Toeplitz reflexive or perfect if $E^{\times\times} = E$. It is to be noted that $E^\times$ is perfect for any sequence space E.

For the analogues in the classical case of the results proved in this section, one can refer to [3], [2]. Many of the results proved in this section point out a significant departure from their analogues in the classical case. For results proved in this section, see [9].

By the observation made in Section 6.2, we have the following result.

Theorem 74. $c_0(p)$ *and $\ell_\infty(p)$ are perfect for all p and $c(p)$ is not perfect for any p.*

Theorem 75. $\ell(p) \neq c_0(p)$ *for any p.*

Proof. It is clear that $\ell(p) \subset c_0(p)$. To prove the theorem, we prove that there exists $\{x_k\} \in c_0(p)$, $\{x_k\} \notin \ell(p)$. Since K is non-trivially valued, there exists $\pi \in K$ such that $0 < \rho = |\pi| < 1$. Now, for $i = 1, 2, \dots$, we can find an integer α_i such that

$$\rho^{\alpha_i + 1} \leq i^{-\frac{1}{p_i}} < \rho^{\alpha_i}.$$

Define $x = \{x_k\}$, where $x_0 = 0$, $x_k = \pi^{\alpha_k}$, $k = 1, 2, \ldots$. Then

$$\sum_{k=0}^{\infty} |x_k|^{p_k} = \sum_{k=1}^{\infty} |x_k|^{p_k}$$

$$= \sum_{k=1}^{\infty} \rho^{\alpha_k p_k}$$

$$> \sum_{k=1}^{\infty} \frac{1}{k}$$

$$= \infty,$$

so that $x = \{x_k\} \notin \ell(p)$. However,

$$|x_k|^{p_k} = \rho^{\alpha_k p_k}$$

$$\leq \frac{1}{k} \frac{1}{\rho^{p_k}}$$

$$< \frac{1}{k} \frac{1}{\rho^H}$$

$$\to 0, \ k \to \infty,$$

where $p_k \leq H$, $k = 0, 1, 2, \ldots$. Consequently, $x = \{x_k\} \in c_0(p)$, completing the proof. $\qquad \square$

In view of Theorem 75, we have:

Theorem 76. *$\ell(p)$ is not perfect for any p.*

Proof. If $\ell(p)$ is perfect for some $p = \{p_k\}$, then

$$\ell(p) = \ell^{\times\times}(p)$$

$$= \ell_\infty^\times(p)$$

$$= c_0(p),$$

which contradicts Theorem 75. $\qquad \square$

We now prove a result bringing out the relationship between the sequence spaces $c_0(p)$ and ℓ_1.

Theorem 77.   *(1)* $\ell_1 \subset c_0(p)$ *if and only if* $\inf\limits_{k \geq 0} p_k > 0$;

(2) $c_0(p) \subset \ell_1$ *if and only if* $p \in Q$;

and

(3) $c_0(p) \neq \ell_1$ *for any* p,

where Q is the set of all $p = \{p_k\}$ for which there exists $N > 1$ such that

$$\sum_{k=0}^{\infty} N^{-\frac{1}{p_k}} < \infty.$$

Proof. (1) If $\inf\limits_{k \geq 0} p_k > 0$, $c_0(p) = c_0$ and so $\ell_1 \subset c_0 = c_0(p)$. On the other hand, if $\ell_1 \subset c_0(p)$, then $\ell_1^\times \supset c_0^\times(p)$, i.e., $\ell_\infty \supset \ell_\infty(p)$, which implies that $\ell_\infty(p) = \ell_\infty$. It now follows that $\inf\limits_{k \geq 0} p_k > 0$, for, otherwise, we can find a strictly increasing sequence $\{k(i)\}$ of positive integers such that

$$p_{k(i)} < \frac{1}{i}, \ i = 1, 2, \dots.$$

Consider the sequence $x = \{x_k\}$, where

$$x_k = \pi^{-\frac{1}{p_{k(i)}}}, \ \ \text{if } k = k(i);$$

$$= 0, \ \text{if } k \neq k(i), \ i = 1, 2, \dots,$$

where $0 < \rho = |\pi| < 1$, $\pi \in K$. Now,

$$|x_k| = \frac{1}{\rho^{\frac{1}{p_{k(i)}}}} \geq \frac{1}{\rho^i} \to \infty, \ i \to \infty,$$

noting that $\frac{1}{\rho} > 1$ and $\frac{1}{p_{k(i)}} > i$. Consequently, $x = \{x_k\} \notin \ell_\infty$. However,

$$|x_{k(i)}|^{p_{k(i)}} \leq \frac{1}{\rho}, \ i = 1, 2, \dots,$$

so that $x = \{x_k\} \in \ell_\infty(p)$, which is a contradiction. Our claim is now established.

(2) Let $p = \{p_k\} \in Q$ and $x = \{x_k\} \in c_0(p)$. Then there exist $N > 1$ and $k_0 = k_0(x, N)$ such that

$$\sum_{k=0}^{\infty} N^{-\frac{1}{p_k}} < \infty \text{ and } |x_k|^{p_k} < \frac{1}{N}, \ k \geq k_0.$$

It now follows that $x = \{x_k\} \in \ell_1$. To prove the converse, let $c_0(p) \subset \ell_1$. Suppose $p \notin Q$, i.e.,

$$\sum_{k=0}^{\infty} N^{-\frac{1}{p_k}} = \infty \text{ for every } N > 1.$$

We can now find a strictly increasing sequence $\{k(N)\}$ of positive integers such that

$$\sum_{k(N-1)<k\leq k(N)} (N+1)^{-\frac{1}{p_k}} > N.$$

Now, given positive integers k and $N > 1$, we can choose an integer $\alpha_N^{(k)}$ such that

$$\rho^{\alpha_N^{(k)}+1} \leq (N+1)^{-\frac{1}{p_k}} < \rho^{\alpha_N^{(k)}},$$

where $0 < \rho = |\pi| < 1$, $\pi \in K$. Consider the sequence $x = \{x_k\}$ defined by

$$x_k = \pi^{\alpha_N^{(k)}}, \ k(N-1) < k \leq k(N), N = 1, 2, \ldots.$$

As before,

$$|x_k|^{p_k} = \rho^{\alpha_N^{(k)} p_k} \leq \frac{1}{N+1} \cdot \frac{1}{\rho^H} \to 0, N \to \infty,$$

where $p_k \leq H$, $k = 0, 1, 2, \ldots$, so that $x = \{x_k\} \in c_0(p)$. However,

$$\sum_{k=0}^{\infty} |x_k| = \sum_{N=1}^{\infty} \sum_{k(N-1)<k\leq k(N)} \rho^{\alpha_N^{(k)}}$$

$$= \sum_{N=1}^{\infty} \sum_{k(N-1)<k\leq k(N)} (N+1)^{-\frac{1}{p_k}}$$

$$> \sum_{N=1}^{\infty} N$$

$$= \infty,$$

and so $x = \{x_k\} \notin \ell_1$, which is a contradiction.

(3) If $p \notin Q$, then in view of (2), $c_0(p) \not\subset \ell_1$ and so $c_0(p) \neq \ell_1$. On the other hand, if $p \in Q$, there exists $N > 1$ such that $\sum_{k=0}^{\infty} N^{-\frac{1}{p_k}} < \infty$. For $k = 1, 2, \ldots$, there exists an integer α_k such that

$$\rho^{\alpha_k+1} \leq N^{-\frac{1}{p_k}} < \rho^{\alpha_k},$$

where, as usual, $0 < \rho = |\pi| < 1$, $\pi \in K$. Consider the sequence $x = \{x_k\}$, $x_k = \pi^{\alpha_k}$, $k = 1, 2, \ldots$. Now,

$$\sum_{k=0}^{\infty} |x_k| = \sum_{k=0}^{\infty} \rho^{\alpha_k}$$

$$\leq \frac{1}{\rho} \sum_{k=0}^{\infty} N^{-\frac{1}{p_k}}$$

$$< \infty,$$

so that $x = \{x_k\} \in \ell_1$, while

$$|x_k|^{p_k} = \rho^{\alpha_k p_k} > \frac{1}{N}, \quad k = 1, 2, \ldots.$$

Consequently, $x = \{x_k\} \notin c_0(p)$. Thus $c_0(p) \neq \ell_1$. The proof of the theorem is now complete. $\square$

We now prove an interesting characterization.

Theorem 78. *The following statements are equivalent:*

(i) $\inf_{k \geq 0} p_k > 0$;

(ii) $c_0(p) = c_0$;

and

(iii) $\ell_\infty(p) = \ell_\infty$.

Proof. It is clear that (i) implies (ii). If (ii) holds, $c_0^\times(p) = c_0^\times$, i.e., $\ell_\infty(p) = \ell_\infty$ so that (iii) holds. If $\ell_\infty(p) = \ell_\infty$, as in the proof of statement (1) of Theorem 77, it follows that (i) holds, completing the proof of the theorem. $\square$

We recall that if X is a linear topological space, a sequence $\{x_n\}$ in X is said to converge weakly to $x \in X$ if

$$\lim_{n \to \infty} f(x_n) = f(x),$$

for every $f \in X^*$, the continuous dual of X.

We also recall that a linear metric space X is said to have the Schur property if every weakly convergent sequence in X is convergent in the metric of the space. Since convergence in the metric of X, i.e., strong convergence always implies weak convergence to the same limit, X has the Schur property if and only if strong and weak convergence are equivalent, with the same limit (cf. Section 5.1). We know that $\ell(p)$ is a complete, linear topological space under the topology induced by the paranorm

$$g(x) = \left(\sum_{k=0}^{\infty} |x_k|^{p_k} \right)^{\frac{1}{H}}, \quad x = \{x_k\} \in \ell(p)$$

and $c_0(p)$ is a complete, linear topological space under the topology induced by the paranorm

$$g(x) = \sup_{k \geq 0} |x_k|^{\frac{p_k}{H}}, \quad x = \{x_k\} \in c_0(p),$$

where $H = \max(1, \sup_{k \geq 0} p_k)$.

It was proved in Section 5.1 that the linear topological space $c_0(p)$ has the Schur property for all p (Theorem 44).

On the other hand, we have the following result about $\ell(p)$.

Theorem 79. *The linear topological space $\ell(p)$ does not have the Schur property if $\inf\limits_{k \geq 0} p_k > 0$, or equivalently, if $\ell(p)$ has the Schur property, then $\inf\limits_{k \geq 0} p_k = 0$.*

Proof. Let $\inf\limits_{k \geq 0} p_k > 0$. Then $\ell_\infty(p) = \ell_\infty$, in view of Theorem 78. For every $k = 1, 2, \ldots$, choose an integer (in fact, non-negative) α_k such that

$$\rho^{\alpha_k + 1} \leq \frac{1}{k^{\frac{1}{H}}} < \rho^{\alpha_k},$$

where, as usual, $H = \max(1, \sup_{k \geq 0} p_k)$ and $0 < \rho = |\pi| < 1$, $\pi \in K$. Define $x = \{x_k\}$ by

$$x_k = \pi^{\alpha_k}, \quad k = 1, 2, \ldots.$$

Let

$$y^{(n)} = \left\{ \underbrace{x_n, x_n, \ldots, x_n}_{n \text{ times}}, 0, 0, \ldots \right\}, \quad n = 0, 1, 2, \ldots.$$

Then $y^{(n)} \in \ell(p)$, $n = 0, 1, 2, \ldots$. Now,

$$\{g(y^{(n)})\}^H = |x_n|^{p_0} + |x_n|^{p_1} + \cdots + |x_n|^{p_{n-1}}$$
$$> n|x_n|^H$$
$$= n\rho^{\alpha_n H}$$
$$> 1,$$

for sufficiently large n, where we can suppose that $|x_n| < 1$ for sufficiently large n and since $p_k \leq H$, $k = 0, 1, 2, \ldots$, $|x_n|^{p_k} > |x_n|^H$, $k = 0, 1, 2, \ldots$. This shows that

$$y^{(n)} \nrightarrow 0, \ n \to \infty \text{ in the metric of } \ell(p).$$

On the other hand, if $f \in \ell^*(p)$, there exists $\{a_k\} \in \ell_\infty(p) = \ell_\infty$ such that

$$f(y^{(n)}) = \sum_{k=0}^{n-1} x_n a_k,$$

in view of Theorem 62. So

$$|f(y^{(n)})| = \left| x_n \sum_{k=0}^{n-1} a_k \right|$$
$$\leq M|x_n|$$
$$= M\rho^{\alpha_n}$$
$$\leq \frac{M}{\rho} \frac{1}{n^{\frac{1}{H}}}$$
$$\to 0, \ n \to \infty,$$

where $|a_k| \leq M$, $k = 0, 1, 2, \ldots$. Thus $\{y^{(n)}\}$ converges weakly to 0, establishing the theorem. □

Corollary 5. *The sequence spaces ℓ_α, $\alpha > 0$, do not have the Schur property. In particular, the space ℓ_1 does not have the Schur property.*

This is invariance with its analogue in the classical case.

6.4 On the algebras (c, c) and $(\ell_\alpha, \ell_\alpha)$

For results in this section, the reader can refer to [7].

It was proved in [11] that the class (c, c) is a non-Archimedean Banach algebra under the norm

$$\|A\| = \sup_{n,k} |a_{nk}|, \tag{6.15}$$

$A = (a_{nk}) \in (c, c)$ with the usual matrix addition, scalar multiplication and multiplication.

We now state below some more results relating to the Banach algebra (c, c), omitting the proofs which are modeled on those for the real or complex case.

Theorem 80. *The class $(c, c; P)$, as a subset of (c, c), is a closed K-convex semigroup with identity.*

Remark 19. *Note that $(c, c; P)$ is not an algebra since the sum of two elements in $(c, c; P)$ may not be in $(c, c; P)$.*

We now introduce a convolution product following [4, p. 179].

Definition 31. *For $A = (a_{nk})$, $B = (b_{nk})$, define*

$$(A * B)_{nk} = \sum_{i=0}^{k} a_{ni} b_{n,k-i}, \quad n, k = 0, 1, 2, \ldots. \tag{6.16}$$

$A * B = \{(A * B)_{nk}\}$ *is called the convolution product I of A and B.*

Keeping the usual norm structure in the class (c, c), defined by (6.15) and replacing the usual matrix product by the convolution product $*$ as defined in (6.16), we have the following.

Theorem 81. *(c, c) is a commutative Banach algebra with identity under the convolution product $*$. Further, $(c, c; P)$, as a subset of (c, c), is a closed K-convex semigroup with identity.*

We now prove theorems analogous to Theorem 80 and Theorem 81 for the class $(\ell_\alpha, \ell_\alpha)$, $\alpha \geq 1$.

Theorem 82. *The class* $(\ell_\alpha, \ell_\alpha; P)$, *as a subset of* $(\ell_\alpha, \ell_\alpha)$, *where* $\alpha \geq 1$, *is a closed K-convex semigroup with identity, the multiplication being the usual matrix multiplication.*

Proof. Let $\lambda, \mu, \gamma \in K$ such that $\lambda + \mu + \gamma = 1$ and $|\lambda|, |\mu|, |\gamma| \leq 1$. Let $A = (a_{nk})$, $B = (b_{nk})$, $C = (c_{nk}) \in (\ell_\alpha, \ell_\alpha, P)$. Using the details in Section 3.4, $\lambda A + \mu B + \gamma C \in (\ell_\alpha, \ell_\alpha)$. Also,

$$\sum_{n=0}^{\infty} (\lambda a_{nk} + \mu b_{nk} + \gamma c_{nk}) = \lambda + \mu + \gamma = 1, \ k = 0, 1, 2, \ldots,$$

so that $\lambda A + \mu B + \gamma C \in (\ell_\alpha, \ell_\alpha; P)$, using the results noted in the beginning of Section 3.5. So $(\ell_\alpha, \ell_\alpha; P)$ is a K-convex subset of $(\ell_\alpha, \ell_\alpha)$.

Let now, $A = (a_{nk}) \in \overline{(\ell_\alpha, \ell_\alpha; P)}$. Then there exist

$$A^{(m)} = (a_{nk}^{(m)}) \in (\ell_\alpha, \ell_\alpha; P), \ m = 0, 1, 2, \ldots$$

such that

$$\|A^{(m)} - A\| \to 0, \ m \to \infty.$$

So, given $\epsilon > 0$, there exists a positive integer N such that

$$\|A^{(m)} - A\| < \epsilon, \ m \geq N,$$

$$i.e., \sup_{k \geq 0} \left(\sum_{n=0}^{\infty} |a_{nk}^{(m)} - a_{nk}|^\alpha \right)^{\frac{1}{\alpha}} < \epsilon, \ m \geq N. \tag{6.17}$$

Again,

$$\left| \sum_{n=0}^{\infty} a_{nk} - 1 \right|^{\alpha} = \left| \sum_{n=0}^{\infty} a_{nk} - \sum_{n=0}^{\infty} a_{nk}^{(N)} \right|^{\alpha},$$

$$\text{since } A^{(N)} \in (\ell_\alpha, \ell_\alpha; P)$$

$$= \left| \sum_{n=0}^{\infty} (a_{nk} - a_{nk}^{(N)}) \right|^{\alpha}$$

$$\leq \sum_{n=0}^{\infty} |a_{nk} - a_{nk}^{(N)}|^{\alpha}, \quad \text{since the valuation}$$

$$\text{on } K \text{ is non-Archimedean}$$

$$< \epsilon^{\alpha}, \ k = 0, 1, 2, \dots, \text{ using (6.17)}.$$

So

$$\left| \sum_{n=0}^{\infty} a_{nk} - 1 \right| < \epsilon \ \text{ for all } \epsilon > 0, k = 0, 1, 2, \dots$$

and consequently

$$\sum_{n=0}^{\infty} a_{nk} = 1, k = 0, 1, 2, \dots.$$

Thus $A \in (\ell_\alpha, \ell_\alpha; P)$ and $(\ell_\alpha, \ell_\alpha; P)$ is a closed subset of $(\ell_\alpha, \ell_\alpha)$.

To complete the proof of the theorem, it suffices to check closure under matrix product. If $A = (a_{nk})$, $B = (b_{nk}) \in (\ell_\alpha, \ell_\alpha; P)$, it is clear that $AB \in (\ell_\alpha, \ell_\alpha)$. However, $AB \in (\ell_\alpha, \ell_\alpha; P)$ too, since

$$\sum_{n=0}^{\infty} (AB)_{nk} = \sum_{n=0}^{\infty} \left(\sum_{i=0}^{\infty} a_{ni} b_{ik} \right)$$

$$= \sum_{i=0}^{\infty} b_{ik} \left(\sum_{n=0}^{\infty} a_{ni} \right), \quad \text{since convergence is}$$

$$\text{equivalent to unconditional convergence}$$

$$\text{(see [13, p. 133])}$$

$$= \sum_{i=0}^{\infty} b_{ik}, \text{ since } \sum_{n=0}^{\infty} a_{ni} = 1, \ i = 0, 1, 2, \dots$$

$$= 1, \text{ since } \sum_{i=0}^{\infty} b_{ik} = 1, \ k = 0, 1, 2, \dots.$$

The proof of the theorem is now complete. $\qquad\square$

Remark 20. $(\ell_\alpha, \ell_\alpha; P)$ *is not an algebra, since the sum of two elements of* $(\ell_\alpha, \ell_\alpha; P)$ *may not be in* $(\ell_\alpha, \ell_\alpha; P)$.

We now introduce another convolution product.

Definition 32. *For* $A = (a_{nk})$, $B = (b_{nk})$, *define*

$$(A \circ B)_{nk} = \sum_{i=0}^{n} a_{ik} b_{n-i,k}, \ n, k = 0, 1, 2, \ldots. \tag{6.18}$$

$A \circ B = ((A \circ B)_{nk})$ *is called the convolution product II of* A *and* B.

We keep the usual norm structure in $(\ell_\alpha, \ell_\alpha)$ and replace matrix product by the convolution product $\circ$ as defined by (6.18) and establish the following.

Theorem 83. $(\ell_\alpha, \ell_\alpha)$, $\alpha \geq 1$, *is a commutative Banach algebra with identity under the convolution product* $\circ$ *and* $(\ell_\alpha, \ell_\alpha; P)$, *as a subset of* $(\ell_\alpha, \ell_\alpha)$, *is a closed K-convex semigroup with identity.*

Proof. We will prove closure under convolution product $\circ$. Let $A = (a_{nk})$, $B = (b_{nk}) \in (\ell_\alpha, \ell_\alpha)$. Then

$$\sum_{n=0}^{\infty} |(A \circ B)_{nk}|^\alpha = \sum_{n=0}^{\infty} \left| \sum_{i=0}^{n} a_{ik} b_{n-i,k} \right|^\alpha$$

$$\leq \sum_{n=0}^{\infty} \sum_{i=0}^{n} |a_{ik}|^\alpha |b_{n-i,k}|^\alpha, \ \text{since the}$$

$$\text{valuation on } K \text{ is non-Archimedean}$$

$$= \left(\sum_{n=0}^{\infty} |a_{nk}|^\alpha \right) \left(\sum_{n=0}^{\infty} |b_{nk}|^\alpha \right)$$

$$\leq \|A\|^\alpha \|B\|^\alpha, \ k = 0, 1, 2, \ldots,$$

so that $\sup_{k \geq 0} \left(\sum_{n=0}^{\infty} |(A \circ B)_{nk}|^\alpha \right) < \infty$ and so $A \circ B \in (\ell_\alpha, \ell_\alpha)$. Also $\|A \circ B\|^\alpha \leq \|A\|^\alpha \|B\|^\alpha$, which implies $\|A \circ B\| \leq \|A\| \|B\|$.

It is clear that the convolution product $\circ$ is commutative. The identity element is the matrix $E = (e_{nk})$, whose first row consists of 1's and which has 0's elsewhere, i.e., $e_{0k} = 1$, $k = 0, 1, 2, \ldots$; $e_{nk} = 0$, $n = 1, 2, \ldots$, $k = 0, 1, 2, \ldots$.

Note that $\|E\| = 1$. We also note that $E \in (\ell_\alpha, \ell_\alpha; P)$, since $\sum\limits_{n=0}^{\infty} e_{nk} = 1$, $k = 0, 1, 2, \ldots$. It suffices to prove that $(\ell_\alpha, \ell_\alpha; P)$ is closed under the convolution product $\circ$. Now,

$$\sum_{n=0}^{\infty} (A \circ B)_{nk} = \sum_{n=0}^{\infty} \left(\sum_{i=0}^{n} a_{ik} b_{n-i,k} \right)$$
$$= \left(\sum_{n=0}^{\infty} b_{n,k} \right) \left(\sum_{n=0}^{\infty} a_{nk} \right)$$
$$= 1, \ k = 0, 1, 2, \ldots,$$

if $A, B \in (\ell_\alpha, \ell_\alpha; P)$, since in this case, $\sum\limits_{n=0}^{\infty} a_{nk} = \sum\limits_{n=0}^{\infty} b_{nk} = 1$, $k = 0, 1, 2, \ldots$. The proof of the theorem is now complete. $\qquad \square$

We now obtain a Mercerian theorem supplementing an earlier one (Theorem 29).

Theorem 84. *When $K = \mathbb{Q}_p$, the p-adic field for a prime p, if*

$$y_n = x_n + \lambda(p^n x_0 + p^{n-1} x_1 + \cdots + p x_{n-1} + x_n)$$

and $\{y_n\} \in \ell_\alpha$, then $\{x_n\} \in \ell_\alpha$, $\alpha \geq 1$, provided

$$|\lambda|_p < (1 - \rho^\alpha)^{\frac{1}{\alpha}},$$

where $\rho = |p|_p < 1$.

Proof. Since $(\ell_\alpha, \ell_\alpha)$, $\alpha \geq 1$, is a Banach algebra under the convolution product $\circ$, if $\lambda \in \mathbb{Q}_p$ is such that $|\lambda|_p < \frac{1}{\|A\|}$, $A \in (\ell_\alpha, \ell_\alpha)$, then $E - \lambda A$, where E is the identity element of $(\ell_\alpha, \ell_\alpha)$ under the convolution product $\circ$, has an inverse in $(\ell_\alpha, \ell_\alpha)$. In this context, we recall that

$$E \equiv (e_{nk}) = \begin{bmatrix} 1 & 1 & 1 & \cdots \\ 0 & 0 & 0 & \cdots \\ 0 & 0 & 0 & \cdots \\ \cdots & \cdots & \cdots & \cdots \end{bmatrix}.$$

We note that the equations

$$y_n = x_n + \lambda(p^n x_0 + p^{n-1} x_1 + \cdots + p x_{n-1} + x_n), \ n = 0, 1, 2, \ldots$$

are given by

$$(E + \lambda A) \circ x' = y',$$

where

$$A = \begin{bmatrix} 1 & 0 & 0 & \cdots \\ p & 0 & 0 & \cdots \\ p^2 & 0 & 0 & \cdots \\ \cdots & \cdots & \cdots & \cdots \end{bmatrix},$$

$$x' = \begin{bmatrix} x_0 & 0 & 0 & \cdots \\ x_1 & 0 & 0 & \cdots \\ x_2 & 0 & 0 & \cdots \\ \cdots & \cdots & \cdots & \cdots \end{bmatrix},$$

and

$$y' = \begin{bmatrix} y_0 & 0 & 0 & \cdots \\ y_1 & 0 & 0 & \cdots \\ y_2 & 0 & 0 & \cdots \\ \cdots & \cdots & \cdots & \cdots \end{bmatrix}.$$

It is clear that $A \in (\ell_\alpha, \ell_\alpha)$ with $\|A\| = \frac{1}{(1-\rho^\alpha)^{\frac{1}{\alpha}}}$, where $\rho = |p|_p < 1$. So, if

$$|\lambda|_p < (1 - \rho^\alpha)^{\frac{1}{\alpha}},$$

$E + \lambda A$ has an inverse in $(\ell_\alpha, \ell_\alpha)$. Consequently, it follows that

$$x' = (E + \lambda A)^{-1} \circ y'.$$

Since $y' \in (\ell_\alpha, \ell_\alpha)$ and $(E + \lambda A)^{-1} \in (\ell_\alpha, \ell_\alpha)$, $x' \in (\ell_\alpha, \ell_\alpha)$ and so $\{x_n\} \in \ell_\alpha$, completing the proof of the theorem. $\qquad \square$

The classical analogues of Theorem 82, Theorem 83 and Theorem 84 for the case $\alpha = 1$ can be found in [8].

Bibliography

[1] R. Bhaskaran and P.N. Natarajan. *Rocky Mountain J. Math.*, 16:129–136, 1986.

[2] C.G. Lascarides. A study of certain spaces of Maddox and a generalization of a theorem of Iyer. *Pacific J. Math.*, 38:487–500, 1971.

[3] I.J. Maddox. Continuous and Köthe-Toeplitz duals of certain sequence spaces. *Proc. Cambridge Philos. Soc.*, 65:431–435, 1969.

[4] I.J. Maddox. *Elements of functional analysis*. Cambridge, 1977.

[5] I.J. Maddox and M.A.L. Willey. Continuous operators on paranormed spaces and matrix transformations. *Pacific J. Math.*, 53:217–228, 1974.

[6] P.N. Natarajan. Continuous duals of certain sequence spaces and the related matrix transformations over non-archimedean fields. *Indian J. Pure Applied Math.*, 21:82–87, 1990.

[7] P.N. Natarajan. On the algebras (c, c) and $(\ell_\alpha, \ell_\alpha)$ in non-archimedean fields. *Proceedings of the International Conference in p-adic Functional Analysis held at Poznan*, Poland, Marcel Dekker, 225–231, 1999.

[8] P.N. Natarajan. On the algebra (ℓ_1, ℓ_1) of infinite matrices. *Analysis (München)*, 20:353–357, 2000.

[9] P.N. Natarajan. Some properties of certain sequence spaces over non-archimedean fields. *Proceedings of the International Conference in p-adic Functional Analysis held at Ionnina*, Greece, Marcel Dekker, 227–232, 2001.

[10] P.N. Natarajan and M.S. Rangachari. Matrix transformations between sequence spaces over non-Archimedean fields. *Rev. Roum. Math. Pures Appl.*, 24:615–618, 1979.

[11] M.S. Rangachari and V.K. Srinivasan. Matrix transformations in non-archimedean fields. *Indag. Math.*, 26:422–429, 1964.

[12] S. Simons. The sequence spaces l(p_v) and m(p_v). *Proc. London Math. Soc.*, 15:422–436, 1965.

[13] A.C.M. Van Rooij and W.H. Schikhof. Non-Archimedean analysis. *Nieuw Arch. Wisk.*, 19:120–160, 1971.

Chapter 7

A Characterization of the Matrix Class (ℓ_∞, c_0) and Summability Matrices of Type M in Non-Archimedean Analysis

7.1 Introduction

Throughout this chapter, K denotes a complete, non-trivially valued, non-Archimedean field. We first prove a Steinhaus-type result. We then obtain a characterization of the matrix class (ℓ_∞, c_0). These results lead us to a characterization of a null sequence among bounded sequences with non-zero entries. For the classical analogues, one can refer to [8]. Summability matrices of type M in K were first introduced by Rangachari and Srinivasan [12]. However, their definition does not seem to be suitable in the non-Archimedean setup. We introduce a more suitable definition of summability matrices of type M in K and make a detailed study of such matrices.

7.2   A Steinhaus-type theorem

We write $A = (a_{nk}) \in (c_0, c_0; P)$ if $A \in (c_0, c_0)$ and $\sum_{n=0}^{\infty}(Ax)_n = \sum_{k=0}^{\infty} x_k$, $x = \{x_k\} \in c_0$.

One can easily prove the following result.

Theorem 85. $A = (a_{nk}) \in (c_0, c_0; P)$ *if and only if*

$$
\left.
\begin{aligned}
&(i) \sup_{n,k} |a_{nk}| < \infty; \\[2mm]
&(ii) a_{nk} \to 0, \; n \to \infty, k = 0, 1, 2, \ldots; \\[2mm]
&and \\[2mm]
&(iii) \sum_{n=0}^{\infty} a_{nk} = 1, k = 0, 1, 2, \ldots.
\end{aligned}
\right\}
\tag{7.1}
$$

It is known [6] that $A = (a_{nk}) \in (\ell_\infty, c_0)$ if and only if

$$
\left.
\begin{aligned}
&(i) a_{nk} \to 0, \; k \to \infty, n = 0, 1, 2, \ldots; \\[2mm]
&and \\[2mm]
&(ii) \sup_{k \geq 0} |a_{nk}| \to 0, \; n \to \infty.
\end{aligned}
\right\}
\tag{7.2}
$$

In this context, it is worth proving that (7.2) is equivalent to

$$
\left.
\begin{aligned}
&(i) a_{nk} \to 0, \; n \to \infty, k = 0, 1, 2, \ldots; \\[2mm]
&and \\[2mm]
&(ii) \sup_{n \geq 0} |a_{nk}| \to 0, \; k \to \infty.
\end{aligned}
\right\}
\tag{7.3}
$$

For, suppose (7.2) holds. For every $k = 0, 1, 2, \ldots$,

$$
|a_{nk}| \leq \sup_{k' \geq 0} |a_{nk'}| \to 0, \; n \to \infty,
$$

using (7.2) (ii). Thus (7.3) (i) holds. Again, by (7.2) (ii), given $\epsilon > 0$, there exists a positive integer N such that

$$\sup_{k \geq 0} |a_{nk}| < \epsilon, \ n > N. \tag{7.4}$$

By (7.2) (i), for $n = 0, 1, 2, \ldots, N$, there exists a positive integer L such that

$$|a_{nk}| < \epsilon, \ k > L. \tag{7.5}$$

(7.4) and (7.5) imply that

$$|a_{nk}| < \epsilon, \ k > L \text{ for all } n = 0, 1, 2, \ldots,$$

i.e., $\sup_{n \geq 0} |a_{nk}| \leq \epsilon, \ k > L,$

i.e., $\sup_{n \geq 0} |a_{nk}| \to 0, \ k \to \infty,$

so that (7.3) (ii) holds. In other words, (7.2) implies (7.3). We can similarly prove that (7.3) implies (7.2). Consequently, $A \in (\ell_\infty, c_0)$ if and only if (7.3) holds.

We now prove a Steinhaus-type result.

Theorem 86. $(c_0, c_0; P) \cap (\ell_\infty, c_0) = \phi.$

Proof. Suppose $A = (a_{nk}) \in (c_0, c_0; P) \cap (\ell_\infty, c_0)$. Then,

$$1 = \left| \sum_{n=0}^{\infty} a_{nk} \right|$$

$$\leq \sup_{n \geq 0} |a_{nk}|$$

$$\to 0, \ k \to \infty,$$

using (7.3) (ii), which is a contradiction. This proves the theorem. $\qquad\square$

7.3   A characterization of the matrix class (ℓ_∞, c_0)

When $K = \mathbb{R}$ or $\mathbb{C}$, Maddox [5] proved a characterization of Schur matrices and its non-Archimedean analogue was proved in [7] (see Theorem 4.1.2).

Similar to the result of Maddox, Natarajan [8] proved a characterization of the matrix class (ℓ_∞, c_0), when $K = \mathbb{R}$ or $\mathbb{C}$. In this section, we prove its analogue in the non-Archimedean case.

Definition 33. *Let $\{\alpha(p)\}$, $\{p(i)\}$ be subsequences of positive integers. Define the sequence $\{y_p\}$ by*

$$\begin{aligned} y_p &= x_{\alpha(p)}, \quad p = p(i), i = 1, 2, \ldots; \\ &= 0, \qquad otherwise. \end{aligned}$$

$\{y_p\}$ *is then obtained by interpolating zeros in a subsequence of the given sequence x. The pattern of interpolation is determined by the subsequences $\{\alpha(p)\}$ and $\{p(i)\}$ of positive integers.*

The class of all zero interpolated subsequences of the sequence x is denoted by $\mathcal{Z}(x)$. If $\{\alpha(p)\}$ and $\{p(i)\}$ are the sequence of all positive integers, then $\{y_p\}$ is the sequence x itself. If $\{p(i)\}$ is the sequence of all positive integers and $\{\alpha(p)\}$ is any subsequence of positive integers, then $\{y_p\}$ is the subsequence $\{x_{\alpha(p)}\}$ of x. Thus the class $\mathcal{Z}(x)$ includes the class of all subsequences of x. It is clear that a sequence $x = \{x_k\}$ converges to r if and only if every sequence in $\mathcal{Z}(\{x_k - r\})$ converges to 0.

Theorem 87. $A = (a_{nk}) \in (\ell_\infty, c_0)$ *if and only if there exists a bounded, non-null sequence $x = \{x_k\}$ with non-zero entries such that each sequence in $\mathcal{Z}(x)$ is transformed by A into a null sequence.*

Proof. If $A \in (\ell_\infty, c_0)$, then the conclusion of the theorem is clear. Conversely, let $x = \{x_k\}$ be a bounded, non-null sequence with non-zero entries such that each sequence in $\mathcal{Z}(x)$ is transformed by A into a null sequence. Since x is a non-null sequence, there exists $\epsilon' > 0$ and a strictly increasing sequence $\{\ell(j)\}$ of positive integers for which

$$|x_{\ell(j)}| > \epsilon', \; j = 1, 2, \ldots. \tag{7.6}$$

We may assume that $|x_k| \leq 1$, $k = 0, 1, 2, \ldots$. First we prove that

$$a_{nk} \to 0, \; k \to \infty, n = 0, 1, 2, \ldots.$$

For, otherwise, there exists $\epsilon'' > 0$, a positive integer m and a strictly increasing sequence $\{k(i)\}$ of positive integers such that

$$|a_{m,k(i)}| > \epsilon'', \quad i = 1, 2, \ldots . \tag{7.7}$$

By (7.6) and (7.7),

$$|a_{m,k(i)} x_{\ell(k(i))}| > \epsilon^2,$$

where $\epsilon = \min(\epsilon', \epsilon'')$. Then the A-transform of the sequence in $\mathscr{Z}(x)$, determined by the subsequences $\{\ell(j)\}$ and $\{k(i)\}$ does not exist, contradicting the hypothesis. Thus

$$a_{nk} \to 0, \quad k \to \infty, n = 0, 1, 2, \ldots .$$

Next we prove that

$$a_{nk} \to 0, \quad n \to \infty, k = 0, 1, 2, \ldots .$$

Consider the sequences $x = \{x_0, x_1, \ldots, x_{k-1}, x_k, x_{k+1}, \ldots\}$ and $\{x_0, x_1, \ldots, x_{k-1}, 0, 0, \ldots\}$ which are in $\mathscr{Z}(x)$. Both of these are transformed by A into null sequences so that their difference is transformed by A into a null sequence. Thus

$$\lim_{n \to \infty} a_{nk} x_k = 0.$$

Since $x_k \neq 0$,

$$\lim_{n \to \infty} a_{nk} = 0, \quad k = 0, 1, 2, \ldots .$$

We next prove that (7.2) (ii) holds. Suppose not. Then for some $\epsilon''' > 0$ and a strictly increasing subsequence $\{n(i)\}$ of positive integers

$$\sup_{k \geq 0} |a_{n(i),k}| > \epsilon''', \quad i = 1, 2, \ldots . \tag{7.8}$$

Let $\epsilon = \min(\epsilon', \epsilon''')$. We may assume that $\epsilon < 1$. Then (7.6) and (7.8) hold with ϵ', ϵ''' replaced by ϵ. Using the fact that $\lim_{n \to \infty} a_{nk} = 0, \ k = 0, 1, 2, \ldots$ and (7.8), we can choose a strictly increasing sequence $\{p(n(i))\}$ of positive integers such that

$$
\left.
\begin{array}{l}
(i) \quad \sup_{0 \le k \le p(n(i-1))} |a_{n(i),k}| < \dfrac{\epsilon^2}{3}; \\[2ex]
and \\[2ex]
(ii)\, |a_{n(i),p(n(i))}| > \epsilon.
\end{array}
\right\}
\tag{7.9}
$$

and using the fact that $a_{nk} \to 0$, $k \to \infty$, $n = 0, 1, 2, \ldots$,

$$
\sup_{k \ge p(n(i+1))} |a_{n(i),k}| < \frac{\epsilon^2}{3}.
\tag{7.10}
$$

Define the sequence $\{y_p\}$ by

$$
\begin{aligned}
y_p &= x_{\ell(p)}, \quad p = p(n(i)), i = 1, 2, \ldots; \\
&= 0, \qquad \text{otherwise.}
\end{aligned}
$$

Then $\{y_p\} \in \mathcal{Z}(x)$ and is determined by the subsequences $\{\ell(j)\}$ and $\{p(n(i))\}$. Now,

$$
\begin{aligned}
\left| \sum_{p=0}^{\infty} a_{n(i),p} y_p \right| &= \left| \sum_{p=0}^{p(n(i-1))} a_{n(i),p} y_p + \sum_{p=p(n(i-1))+1}^{p(n(i+1))-1} a_{n(i),p} y_p \right. \\
&\qquad \left. + \sum_{p \ge p(n(i+1))} a_{n(i),p} y_p \right| \\
&= \left| \sum_{p=0}^{p(n(i-1))} a_{n(i),p} y_p + a_{n(i),p(n(i))} x_{\ell(p(n(i)))} \right. \\
&\qquad \left. + \sum_{p \ge p(n(i+1))} a_{n(i),p} y_p \right| \\
&\ge |a_{n(i),p(n(i))}| |x_{\ell(p(n(i)))}| \\
&\qquad - \sup_{0 \le p \le p(n(i-1))} |a_{n(i),p}| \\
&\qquad - \sup_{p \ge p(n(i+1))} |a_{n(i),p}| \\
&> \epsilon^2 - \frac{\epsilon^2}{3} - \frac{\epsilon^2}{3}, \quad \text{using } (7.6), (7.9), (7.10) \\
&= \frac{\epsilon^2}{3}, \quad i = 1, 2, \ldots.
\end{aligned}
$$

Hence $\{y_p\} \in \mathcal{Z}(x)$ is not transformed by A into a null sequence, which is a contradiction. Thus $A \in (\ell_\infty, c_0)$, completing the proof of the theorem. $\square$

As an immediate consequence of Theorem 87, we have a result like Buck's theorem [3].

Theorem 88. *A bounded sequence $x = \{x_k\}$ with non-zero entries is a null sequence if and only if there exists a matrix $A \in (c_0, c_0; P)$, which transforms every sequence in $\mathcal{Z}(x)$ into a null sequence.*

Proof. If x is a null sequence, then every sequence in $\mathcal{Z}(x)$ is also a null sequence so that any $A \in (c_0, c_0; P)$ transforms every sequence in $\mathcal{Z}(x)$ into a null sequence. Conversely, let there exist a matrix $A \in (c_0, c_0; P)$ which transforms every sequence in $\mathcal{Z}(x)$ into a null sequence. We claim that x is a null sequence. If not, it follows from Theorem 87 that $A \in (\ell_\infty, c_0)$, contradicting Theorem 86. This completes the proof. $\square$

7.4   Summability matrices of type M

Summability matrices of type M in K were first introduced by Rangachari and Srinivasan [12]. However, their definition does not seem to be suitable in the non-Archimedean setup. The space c_0 in K seems to replace the classical space ℓ_1, since there are several instances when absolute convergence in classical analysis is replaced effectively by usual convergence in a complete, non-trivially valued, non-Archimedean field K. Hence the following definition seems to be more suited for the non-Archimedean case (see [10] for details).

Definition 34. *An infinite matrix $A = (a_{nk})$, $a_{nk} \in K$, $n, k = 0, 1, 2, \ldots$ is said to be of type M if*

$$\{c_n\} \in \ell_\infty \ \text{ and } \ \sum_{i=0}^{\infty} c_i a_{ik} = 0, \ \ k = 0, 1, 2, \ldots \tag{7.11}$$

together imply that

$$c_n = 0, \ n = 0, 1, 2, \ldots, \tag{7.12}$$

where, as usual, ℓ_∞ denotes the set of all bounded sequences in K.

Theorem 89. *A normal, regular method $A = (a_{nk})$ is of type M if*

$$\sup_n |a'_{nk}| < \infty, \ k = 0, 1, 2, \ldots,$$

where $A^{-1} = (a'_{nk})$ is the inverse of A.

Proof. $\{c_n\} \in \ell_\infty$, $\displaystyle\sum_{n=\gamma}^{\infty} c_n a_{n\gamma} = 0$ imply

$$\begin{aligned}
0 &= \sum_{\gamma=k}^{\infty} a'_{\gamma k} \sum_{n=\gamma}^{\infty} c_n a_{n\gamma} \\
&= \sum_{n=k}^{\infty} c_n \sum_{\gamma=k}^{\infty} a_{n\gamma} a'_{\gamma k} \\
&= c_k, \ k = 0, 1, 2, \ldots,
\end{aligned}$$

proving that A is of type M.   □

Example 6. *All weighted means (see [9]) are of type M, since $a'_{n\gamma} = 0$, $\gamma \le n - 2$.*

Let c, c_0, as usual, respectively denote the space of all convergent sequences and the space of all null sequences in K. If A is any convergence-preserving matrix, i.e., $A \in (c, c)$, by the range $R_c(A)$, we mean the set of all $y \in c$ such that $y = Ax$ for some $x \in c$. It is clear that $R_c(A) \subset c$.

Since the Hahn-Banach theorem is available in the non-Archimedean context too if K is spherically complete (see [1, p. 78]), we can prove the following theorem, the proof of which is similar to that of the classical case (see [4]).

Theorem 90. *A reversible regular matrix $A = (a_{nk})$ is of type M if and only if $R_c(A)$ is dense in c, when K is a non-trivially valued, non-Archimedean field which is spherically complete.*

Note that the p-adic field $\mathbb{Q}_p$, for a prime p, is spherically complete and so Theorem 90 holds in such fields.

Definition 35. *Let X be a normed linear space. A subset G of X is said to be "total" in X if every continuous linear functional on X which vanishes for every $x \in G$, vanishes for every $x \in X$. The subset G of X is said to be "fundamental" in X if the linear subspace of X generated by G is dense in X.*

The following result is true in the non-Archimedean setup too (for the analogue in the classical case, see [2, p. 58]), the proof of which is similar to that of the classical case.

Theorem 91. *The subset G of a normed linear space X is fundamental in X if and only if it is total.*

To justify the introduction of a new definition of summability matrices of type M in the non-Archimedean setup, we prove some characterizations of regular matrices of type M, though the proofs of these results are similar to their classical analogues (see [11, Theorem I, Theorem II]).

Theorem 92. *A regular matrix A is of type M if and only if the columns of A are fundamental in c_0.*

Proof. Since A is regular, $\lim_{n \to \infty} a_{nk} = 0$, $k = 0, 1, 2, \ldots$ and so the columns of A are elements of c_0. Recall that any continuous linear functional on c_0 is of the form $\sum_{i=0}^{\infty} c_i x_i$, where $\{x_i\} \in c_0$ and $\{c_i\} \in \ell_\infty$. The values of this functional on the columns of A are $\sum_{i=0}^{\infty} c_i a_{ik}$, $\{c_i\} \in \ell_\infty$, $k = 0, 1, 2, \ldots$. Suppose A is of type M. Then $\sum_{i=0}^{\infty} c_i a_{ik} = 0$, $\{c_i\} \in \ell_\infty$, $k = 0, 1, 2, \ldots$ will imply that $c_n = 0$, $n = 0, 1, 2, \ldots$ so that the value of the functional on the whole space c_0 is zero. Thus the columns of A are total in c_0 and so they are fundamental in c_0, in view of Theorem 91. Conversely, let the columns of A be fundamental in c_0. So, they are total in c_0, again using Theorem 91. Consequently, any continuous linear functional which vanishes on the columns of A, vanishes on c_0. In particular, it vanishes on e_n, $n = 0, 1, 2, \ldots$, where $e_n = \{0, 0, \ldots, 0, 1, 0, \ldots\}$, 1 occurring in the nth place. Hence $c_n = 0$, $n = 0, 1, 2, \ldots$. So A is of type M, completing the proof.   □

Theorem 93. *A regular matrix A is type M if and only if $R_{c_0}(A)$ is fundamental in c_0.*

Proof. First we note that $R_{c_0}(A) \subset c_0$. The columns of A are the A-transforms of the sequences e_n, $n = 0, 1, 2, \ldots$ mentioned in the proof of Theorem 92. Thus the set of all columns of A is a subset of $R_{c_0}(A)$, which is a subset of c_0. Suppose A is of type M. Then the columns of A are fundamental in c_0. It now follows that $R_{c_0}(A)$ is fundamental in c_0.

Conversely, let $R_{c_0}(A)$ be fundamental in c_0. To prove that A is of type M, it suffices to prove that the columns of A are fundamental in c_0, in view of Theorem 92. The values of any continuous linear functional on c_0 over the columns of A are $\sum_{i=0}^{\infty} c_i a_{ik}$, $k = 0, 1, 2, \ldots$, $\{c_i\} \in \ell_\infty$. Its values on the elements of $R_{c_0}(A)$ are $\sum_{i=0}^{\infty} c_i A_i(x)$, $A_i(x) = \sum_{k=0}^{\infty} a_{ik} x_k$, $\{x_k\} \in c_0$, $\{c_i\} \in \ell_\infty$. Suppose $\sum_{i=0}^{\infty} c_i a_{ik} = 0$, $k = 0, 1, 2, \ldots$. Then

$$\sum_{i=0}^{\infty} c_i A_i(x) = \sum_{i=0}^{\infty} c_i \left(\sum_{k=0}^{\infty} a_{ik} x_k \right)$$

$$= \sum_{k=0}^{\infty} x_k \left(\sum_{i=0}^{\infty} c_i a_{ik} \right), \quad \text{since convergence is}$$

equivalent to unconditional convergence

(see [13, p. 133])

$$= 0.$$

Thus the columns of A are fundamental in $R_{c_0}(A)$. Since $R_{c_0}(A)$ is fundamental in c_0, it now follows that the columns of A are fundamental in c_0. The proof of the theorem is now complete.   □

Following [11], we can prove the following results too.

Theorem 94. *The product of two regular matrices of type M is again a regular matrix of type M.*

Theorem 95. *If the product AB of two regular matrices A, B is of type M, then A is of type M.*

We write $A = (a_{nk}) \in (c_0, c; P')$ if $A \in (c_0, c)$ and $\displaystyle\lim_{n\to\infty} A_n(x) = \sum_{k=0}^{\infty} x_k$, $x = \{x_k\} \in c_0$. It is known (see [10]) that $A \in (c_0, c; P')$ if and only if

$$\left.\begin{array}{l} (i)\ \sup_{n,k} |a_{nk}| < \infty; \\[2mm] and \\[2mm] (ii)\ \lim_{n\to\infty} a_{nk} = 1, k = 0, 1, 2, \ldots . \end{array}\right\} \tag{7.13}$$

A is called a γ-matrix if $A \in (c_0, c; P')$. A is called an α-matrix if $A \in (c_0, c_0; P)$.

We now prove the non-Archimedean analogues of certain results of Vermes [14, 15].

Theorem 96. *The product of a regular matrix A and a γ-matrix B is a γ-matrix.*

Proof. Let $A = (a_{nk})$ be regular and $B = (b_{nk})$ be a γ-matrix. Let $AB = (c_{nk})$. It is clear that $\sup_{n,k} |c_{nk}| < \infty$. For $k = 0, 1, 2, \ldots,$

$$\lim_{n\to\infty} c_{nk} - 1 = \lim_{n\to\infty} \sum_{i=0}^{\infty} a_{ni} b_{ik} - \lim_{n\to\infty} \sum_{i=0}^{\infty} a_{ni}$$

$$= \lim_{n\to\infty} \sum_{i=0}^{\infty} a_{ni}(b_{ik} - 1). \tag{7.14}$$

Since $\lim_{i\to\infty} b_{ik} = 1$, given $\epsilon > 0$, there exists a positive integer i_0 such that

$$|b_{ik} - 1| < \frac{\epsilon}{M_1}, \ i > i_0, \tag{7.15}$$

where $M_1 = \sup\limits_{n,k} |a_{nk}| < \infty$. Since $\lim\limits_{n\to\infty} a_{ni} = 0$, $i = 0, 1, 2, \ldots, i_0$, we can choose a positive integer N such that

$$|a_{ni}| < \frac{\epsilon}{M_2}, \ n > N, i = 0, 1, 2, \ldots, i_0, \tag{7.16}$$

where $M_2 = \max\left[1, \sup_{n,k} |b_{nk}|\right] < \infty$. Now,

$$
\left|\sum_{i=0}^{\infty} a_{ni}(b_{ik} - 1)\right| = \left|\sum_{i=0}^{i_0} a_{ni}(b_{ik} - 1) + \sum_{i>i_0} a_{ni}(b_{ik} - 1)\right|
$$

$$
\leq \max\left[\max_{0 \leq i \leq i_0} |a_{ni}||b_{ik} - 1|,\right.
$$

$$
\left. \sup_{i>i_0} |a_{ni}||b_{ik} - 1|\right]
$$

$$
< \max\left[\frac{\epsilon}{M_2}M_2, M_1\frac{\epsilon}{M_1}\right], \ \text{using (7.15)}
$$

$$
\text{and (7.16)}
$$

$$
= \epsilon, \ n > N,
$$

$$i.e., \ \lim_{n\to\infty} \sum_{i=0}^{\infty} a_{ni}(b_{ik} - 1) = 0. \tag{7.17}$$

Using (7.17) in (7.14), we have,

$$\lim_{n\to\infty} c_{nk} = 1, \ k = 0, 1, 2, \ldots.$$

In other words, AB is a γ-matrix, completing the proof. $\qquad\square$

Theorem 97. *The product of a γ-matrix A and an α-matrix B is a γ-matrix.*

Proof. Let $A = (a_{nk})$ be a γ-matrix and $B = (b_{nk})$ be an α-matrix. Let $AB = (c_{nk})$. Note that $\sup\limits_{n,k} |c_{nk}| < \infty$. Now,

$$c_{nk} - 1 = \sum_{i=0}^{\infty} a_{ni}b_{ik} - \sum_{i=0}^{\infty} b_{ik}$$

$$= \sum_{i=0}^{\infty} (a_{ni} - 1)b_{ik}.$$

As in Theorem 96, we can prove that

$$\lim_{n \to \infty} c_{nk} = 1, \ k = 0, 1, 2, \dots,$$

proving that AB is a γ-matrix.   □

Theorem 98. *The product of two α-matrices A, B is again an α-matrix.*

Proof. Let $A = (a_{nk})$, $B = (b_{nk})$ be α-matrices and $AB = (c_{nk})$. It is clear that $\sup_{n,k} |c_{nk}| < \infty$. Now, for $k = 0, 1, 2, \dots,$

$$
\begin{aligned}
\sum_{n=0}^{\infty} c_{nk} &= \sum_{n=0}^{\infty} \left(\sum_{i=0}^{\infty} a_{ni} b_{ik} \right) \\
&= \sum_{i=0}^{\infty} b_{ik} \left(\sum_{n=0}^{\infty} a_{ni} \right), \ \text{interchanging the order of}
\end{aligned}
$$

summation (see [13, p. 133])

$$= \sum_{i=0}^{\infty} b_{ik}, \ \text{since} \ \sum_{n=0}^{\infty} a_{ni} = 1, \ i = 0, 1, 2, \dots$$

$$= 1, \ k = 0, 1, 2, \dots, \ \text{since} \ \sum_{i=0}^{\infty} b_{ik} = 1, \ k = 0, 1, 2, \dots,$$

from which it follows that AB is an α-matrix.   □

We now extend Theorems 96-98 to matrices of type M.

Theorem 99. *If A is a regular matrix of type M and B is a γ-matrix of type M, then their product AB is a γ-matrix of type M.*

Proof. We have already proved that AB is a γ-matrix. Let $AB = (u_{nk})$. Let us suppose that $\sum_{n=0}^{\infty} c_n u_{nk} = 0$, $k = 0, 1, 2, \dots,$ where $\{c_n\} \in \ell_\infty$. Now, for

$k = 0, 1, 2, \ldots,$

$$0 = \sum_{n=0}^{\infty} c_n u_{nk}$$

$$= \sum_{n=0}^{\infty} c_n \left(\sum_{i=0}^{\infty} a_{ni} b_{ik} \right)$$

$$= \sum_{i=0}^{\infty} b_{ik} \left(\sum_{n=0}^{\infty} c_n a_{ni} \right), \text{ interchanging the order of}$$

summation (see [13, p. 133]).

This implies that

$$\sum_{n=0}^{\infty} c_n a_{ni} = 0, \ i = 0, 1, 2, \ldots,$$

since B is of type M. This, in turn, implies that

$$c_n = 0, \ n = 0, 1, 2, \ldots,$$

since A is of type M. Consequently, AB is of type M, which completes the proof. □

We can prove the following results too in a similar fashion.

Theorem 100. *If A is a γ-matrix of type M and B is an α-matrix of type M, then AB is a γ-matrix of type M.*

Theorem 101. *If A, B are α-matrices of type M, then AB is also an α-matrix of type M.*

Bibliography

[1] G. Bachman, E. Beckenstein and L. Narici. *Functional analysis and valuation theory.* Marcel Dekker, 1972.

[2] S. Banach. *Theorie des opérations linéaires.* Warsaw, 1932.

[3] R.C. Buck. A note on subsequences. *Bull. Amer. Math. Soc.*, 49:898–899, 1943.

[4] J.D. Hill. On perfect methods of summability. *Duke Math. J.*, 3:702–714, 1937.

[5] I.J. Maddox. A Tauberian theorem for subsequences. *Bull. London Math. Soc.*, 2:63–65, 1970.

[6] P.N. Natarajan. The Steinhaus theorem for Toeplitz matrices in non-archimedean fields. *Comment. Math. Prace Mat.*, 20:417–422, 1978.

[7] P.N. Natarajan. Characterization of regular and Schur matrices over non-archimedean fields. *Proc. Kon. Ned. Akad. Wetens, Series A*, 90:423–430, 1987.

[8] P.N. Natarajan. A characterization of the matrix class (l_∞, c_0). *Bull. London Math. Soc.*, 23:267–268, 1991.

[9] P.N. Natarajan. Weighted means in non-archimedean fields *Ann. Math. Blaise Pascal*, 2:191–200, 1995.

[10] P.N. Natarajan. Summability matrices of type M in non-archimedean analysis. *African Diaspora J. Math.*, 8:29–35, 2009.

[11] M.S. Ramanujan. On summability methods of type M. *J. London Math. Soc.*, 29:184–189, 1954.

[12] M.S. Rangachari and V.K. Srinivasan. Matrix transformations in non-archimedean fields. *Indag. Math.*, 26:422–429, 1964.

[13] A.C.M. van Rooij and W.H. Schikhof. Non-archimedean analysis. *Nieuw Arch. Wisk.*, 19:120–160, 1971.

[14] P. Vermes. Product of a T-matrix and γ-matrix. *J. London Math. Soc.*, 21:129–134, 1946.

[15] P. Vermes. Series to series transformations and analytic continuation by matrix methods *Amer. J. Math.*, 71:541–562, 1949.

Chapter 8

More Steinhaus-Type Theorems over Valued Fields

8.1 Introduction

In some of the previous chapters, we proved a few Steinhaus-type theorems in valued fields. In this chapter, we prove more Steinhaus-type theorems over valued fields. Throughout this chapter, the field K may be $\mathbb{R}$ or $\mathbb{C}$ or a complete, non-trivially valued, non-Archimedean field. In the relevant context, we explicitly mention which field is chosen. In the absence of such an explicit mention, the field K can be any one of these. For the details in this chapter, the reader can refer to ([4]-[7], [10]). In Sections 8.2-8.5, we present a systematic and detailed study of Steinhaus-type theorems over valued fields. It serves as a good comparative study of Steinhaus-type theorems in the classical case and in the non-Archimedean case.

8.2 A Steinhaus-type theorem when $K = \mathbb{R}$ or $\mathbb{C}$

Let $K = \mathbb{R}$ or $\mathbb{C}$. We need the following sequence space.

$$\gamma = \left\{ x = \{x_k\} : \sum_{k=0}^{\infty} x_k \text{ converges} \right\}.$$

Note that $\ell_1 \subset \gamma \subset c_0 \subset c \subset \ell_\infty$. In this context, we recall the following:

We write $A = (a_{nk}) \in (\ell_1, \ell_1; P)$ if $A \in (\ell_1, \ell_1)$ and $\sum_{n=0}^{\infty} (Ax)_n = \sum_{k=0}^{\infty} x_k$, $x = \{x_k\} \in \ell_1$. We know that $A \in (\ell_1, \ell_1; P)$ if and only if $A \in (\ell_1, \ell_1)$ and $\sum_{n=0}^{\infty} a_{nk} = 1$, $k = 0, 1, 2, \ldots$. We now write

$$A \in (\ell_1, \ell_1; P)'$$

if $A \in (\ell_1, \ell_1; P)$ with $a_{nk} \to 0$, $k \to \infty$, $n = 0, 1, 2, \ldots$. Maddox [2] noted that

$$(\ell_1, \ell_1; P) \cap (\gamma, \ell_1) \neq \phi.$$

In this section, we prove that

$$(\ell_1, \ell_1; P)' \cap (\gamma, \ell_1) = \phi.$$

In particular, a lower triangular $(\ell_1, \ell_1; P)$ matrix cannot belong to the class (γ, ℓ_1).

We now establish the following lemma, the proof of which is modeled on that of Fridy (see [1]).

Lemma 7. *If $A = (a_{nk}) \in (\ell_1, \ell_1)$ with $a_{nk} \to 0$, $k \to \infty$, $n = 0, 1, 2, \ldots$ and*

$$\varlimsup_{k \to \infty} \left| \sum_{n=0}^{\infty} a_{nk} \right| > 0,$$

then there exists a sequence $x = \{x_k\} \in \gamma$ such that $Ax = \{(Ax)_n\} \notin \ell_1$.

Proof. By hypothesis, for some $\epsilon > 0$, there exists a strictly increasing sequence $\{k(i)\}$ of positive integers such that

$$\left| \sum_{n=0}^{\infty} a_{n,k(i)} \right| \geq 2\epsilon, \ i = 1, 2, \ldots.$$

In particular

$$\left| \sum_{n=0}^{\infty} a_{n,k(1)} \right| \geq 2\epsilon.$$

We then choose a positive integer $n(1)$ such that

$$\sum_{n > n(1)} |a_{n,k(1)}| < \min \left(\frac{1}{2}, \frac{\epsilon}{2} \right),$$

this being possible since $\sum\limits_{n=0}^{\infty} |a_{nk}| < \infty$, $k = 0, 1, 2, \ldots$, in view of the fact that $A \in (\ell_1, \ell_1)$. Now, it follows that

$$\left| \sum_{n=0}^{n(1)} a_{n,k(1)} \right| > \epsilon.$$

In general, having chosen $k(j)$, $n(j)$, $j \leq m - 1$, choose a positive integer $k(m) > k(m - 1)$ such that

$$\left| \sum_{n=0}^{\infty} a_{n,k(m)} \right| \geq 2\epsilon,$$

$$\sum_{n=0}^{n(m-1)} |a_{n,k(m)}| < \min\left(\frac{1}{2}, \frac{\epsilon}{2} \right),$$

and then choose a positive integer $n(m) > n(m - 1)$ such that

$$\sum_{n>n(m)} |a_{n,k(m)}| < \min\left(\frac{1}{2^m}, \frac{\epsilon}{2} \right),$$

so that

$$\left| \sum_{n=n(m-1)+1}^{n(m)} a_{n,k(m)} \right| > 2\epsilon - \frac{\epsilon}{2} - \frac{\epsilon}{2}$$

$$= \epsilon.$$

Let the sequence $x = \{x_k\}$ be defined by

$$x_k = \frac{(-1)^{i+1}}{i}, \quad k = k(i);$$

$$= 0, \quad k \neq k(i), i = 1, 2, \ldots.$$

It is clear that $x = \{x_k\} \in \gamma$. Defining $n(0) = 0$, we have,

$$\sum_{n=0}^{n(N)} |(Ax)_n| \geq \sum_{m=1}^{N} \sum_{n=n(m-1)+1}^{n(m)} |(Ax)_n|$$

$$= \sum_{m=1}^{N} \sum_{n=n(m-1)+1}^{n(m)} \left| \sum_{i=1}^{\infty} a_{n,k(i)} x_{k(i)} \right|$$

$$= \sum_{m=1}^{N} \sum_{n=n(m-1)+1}^{n(m)} \left| \sum_{i=1}^{\infty} a_{n,k(i)} \frac{(-1)^{i+1}}{i} \right|$$

$$\geq \sum_{m=1}^{N} \sum_{n=n(m-1)+1}^{n(m)} \left\{ \left| \frac{(-1)^{m+1}}{m} a_{n,k(m)} \right| \right.$$

$$\left. - \sum_{\substack{i=1 \\ i \neq m}}^{\infty} \left| \frac{(-1)^{i+1}}{i} a_{n,k(i)} \right| \right\}$$

$$= \sum_{m=1}^{N} \sum_{n=n(m-1)+1}^{n(m)} \left\{ \frac{1}{m} |a_{n,k(m)}| \right.$$

$$\left. - \sum_{\substack{i=1 \\ i \neq m}}^{\infty} \frac{1}{i} |a_{n,k(i)}| \right\}$$

$$> \epsilon \sum_{m=1}^{N} \frac{1}{m} - \sum_{m=1}^{N} \sum_{n=n(m-1)+1}^{n(m)} \sum_{\substack{i=1 \\ i \neq m}}^{\infty} |a_{n,k(i)}|,$$

$$\text{since } \frac{1}{i} \leq 1.$$

Now,

$$\sum_{m=1}^{\infty} \sum_{n=n(m-1)+1}^{n(m)} \sum_{i<m} |a_{n,k(i)}|$$

$$= \sum_{j=1}^{\infty} \sum_{i>m(j)} |a_{i,k(j)}|$$

$$\leq \sum_{m=1}^{\infty} \frac{1}{2^m}$$

$$= 1.$$

Similarly,

$$\sum_{m=1}^{\infty} \sum_{n=n(m-1)+1}^{n(m)} \sum_{i>m} |a_{n,k(i)}| \leq \frac{1}{2}.$$

So

$$\sum_{n=0}^{n(N)} |(Ax)_n| > \epsilon \sum_{m=1}^{N} \frac{1}{m} - \frac{3}{2}.$$

Since $\sum_{m=1}^{\infty} \frac{1}{m}$ diverges, $Ax = \{(Ax)_n\} \notin \ell_1$, completing the proof of the lemma.

$\square$

We now prove the main result of this section.

Theorem 102. $(\ell_1, \ell_1; P)' \cap (\gamma, \ell_1) = \phi.$

Proof. Let $A = (a_{nk}) \in (\ell_1, \ell_1; P)' \cap (\gamma, \ell_1)$. Since $\sum_{n=0}^{\infty} a_{nk} = 1$, $k = 0, 1, 2, \ldots,$

$$\overline{\lim_{k \to \infty}} \left| \sum_{n=0}^{\infty} a_{nk} \right| = 1 > 0.$$

In view of the preceding Lemma 7, there exists a sequence $x = \{x_k\} \in \gamma$ such that $Ax = \{(Ax)_n\} \notin \ell_1$, which is a contradiction, establishing the theorem.

$\square$

Remark 21. *If the condition $a_{nk} \to 0$, $k \to \infty$, $n = 0, 1, 2, \ldots$ is dropped, the above theorem fails to hold as the following example illustrates. The matrix*

$$A = (a_{nk}) = \begin{bmatrix} 1 & 1 & 1 & \cdots \\ 0 & 0 & 0 & \cdots \\ 0 & 0 & 0 & \cdots \\ \cdots & \cdots & \cdots & \cdots \end{bmatrix}$$

is in $(\ell_1, \ell_1; P) \cap (\gamma, \ell_1)$ and $a_{0k} \nrightarrow 0$, $k \to \infty$, which establishes our claim.

8.3   Some Steinhaus-type theorems over valued fields

In this section, $K = \mathbb{R}$ or $\mathbb{C}$ or a complete, non-trivially valued, non-Archimedean field. We will state explicitly which field is chosen depending on the context.

$(\ell_1, c; P')$ denotes the class of all matrices $A = (a_{nk}) \in (\ell_1, c)$ such that

$$\lim_{n \to \infty} (Ax)_n = \sum_{k=0}^{\infty} x_k, \quad x = \{x_k\} \in \ell_1.$$

When $K = \mathbb{R}$ or $\mathbb{C}$, it is known ([11, p. 4, 17]) that $A = (a_{nk}) \in (\ell_1, c)$ if and only if

$$\left. \begin{array}{l} (i) \ \sup_{n,k} |a_{nk}| < \infty; \\[2mm] and \\[2mm] (ii) \ \lim_{n \to \infty} a_{nk} = \delta_k \text{ exists, } k = 0, 1, 2, \ldots \end{array} \right\} \tag{8.1}$$

We now prove the following.

Theorem 103. *When $K = \mathbb{R}$ or $\mathbb{C}$, $A = (a_{nk}) \in (\ell_1, c; P')$ if and only if (8.1) holds with $\delta_k = 1$, $k = 0, 1, 2, \ldots$.*

Proof. Let $A \in (\ell_1, c; P')$. Let e_k be the sequence $\{0, \ldots, 0, 1, 0, \ldots\}$, 1 occurring in the kth place and 0 elsewhere, $k = 0, 1, 2, \ldots$. Then $e_k \in \ell_1$, $k = 0, 1, 2, \ldots$ and $(Ae_k)_n = a_{nk}$ so that $\lim_{n \to \infty} a_{nk} = 1$, i.e., $\delta_k = 1$, $k = 0, 1, 2, \ldots$ so that (8.1) holds with $\delta_k = 1$, $k = 0, 1, 2, \ldots$.

Conversely, let (8.1) hold with $\delta_k = 1$, $k = 0, 1, 2, \ldots$. Let $x = \{x_k\} \in \ell_1$. In view of (8.1) (i), $\sum_{k=0}^{\infty} a_{nk} x_k$ converges, $n = 0, 1, 2, \ldots$. Now,

$$\begin{aligned} (Ax)_n &= \sum_{k=0}^{\infty} a_{nk} x_k \\ &= \sum_{k=0}^{\infty} (a_{nk} - 1) x_k + \sum_{k=0}^{\infty} x_k, \end{aligned}$$

noting that $\sum\limits_{k=0}^{\infty} a_{nk}x_k$, $\sum\limits_{k=0}^{\infty} x_k$, both converge. Since $\sum\limits_{k=0}^{\infty} |x_k| < \infty$, given $\epsilon > 0$, there exists a positive integer N such that

$$\sum_{k=N+1}^{\infty} |x_k| < \frac{\epsilon}{2A}, \tag{8.2}$$

where $A = \sup\limits_{n,k} |a_{nk} - 1| < \infty$. Since $\lim\limits_{n\to\infty} a_{nk} = 1$, $k = 0, 1, 2, \ldots, N$, a positive integer $N' > N$ can be chosen such that

$$|a_{nk} - 1| < \frac{\epsilon}{2(N+1)M}, \quad n \geq N', k = 0, 1, 2, \ldots, N, \tag{8.3}$$

where, $|x_k| \leq M$, $k = 0, 1, 2, \ldots$, $M > 0$. Now, for $n \geq N'$,

$$\left| \sum_{k=0}^{\infty} (a_{nk} - 1)x_k \right| \leq \sum_{k=0}^{N} |a_{nk} - 1||x_k| + \sum_{k>N} |a_{nk} - 1||x_k|$$

$$< (N+1)\frac{\epsilon}{2(N+1)M}M + A\frac{\epsilon}{2A}, \text{ using (8.2)}$$

and (8.3)

$$= \frac{\epsilon}{2} + \frac{\epsilon}{2}$$

$$= \epsilon,$$

so that

$$\lim_{n\to\infty} \sum_{k=0}^{\infty} (a_{nk} - 1)x_k = 0.$$

Thus

$$\lim_{n\to\infty} (Ax)_n = \sum_{k=0}^{\infty} x_k,$$

$$i.e., A \in (\ell_1, c; P'),$$

which completes the proof of the theorem. □

We now have the following Steinhaus-type result.

Theorem 104. $(\ell_1, c; P') \cap (\ell_p, c) = \phi$, $p > 1$.

Proof. Let $A = (a_{nk}) \in (\ell_1, c; P') \cap (\ell_p, c)$, $p > 1$. It is known ([11, p. 4, 16]) that $A \in (\ell_p, c)$, $p > 1$ if and only if (8.1) (ii) holds and

$$\sup_{n \geq 0} \sum_{k=0}^{\infty} |a_{nk}|^q < \infty, \tag{8.4}$$

where $\frac{1}{p} + \frac{1}{q} = 1$. It now follows that $\sum_{k=0}^{\infty} |\delta_k|^q < \infty$, which contradicts the fact that $\delta_k = 1$, $k = 0, 1, 2, \dots$, since $A \in (\ell_1, c; P')$ and consequently, $\sum_{k=0}^{\infty} |\delta_k|^q$ diverges. This proves our claim. $\square$

Remark 22. *If $K = \mathbb{R}$ or $\mathbb{C}$, since $(\ell_\infty, c) \subset (c, c) \subset (c_0, c) \subset (\ell_p, c)$, $p > 1$, we have,*

$$(\ell_1, c; P') \cap (X, c) = \phi,$$

when $X = \ell_\infty, c, c_0, \ell_p$, $p > 1$.

When K is a complete, non-trivially valued, non-Archimedean field, one can prove that Theorem 103 continues to hold. In this case, if $A = (a_{nk}) \in (\ell_1, c; P') \cap (\ell_\infty, c)$, then

$$\lim_{n \to \infty} \sup_{k \geq 0} |a_{nk} - 1| = 0$$

(see [3, Theorem 2]). So, for any ϵ, $0 < \epsilon < 1$, there exists a positive integer N such that

$$|a_{nk} - 1| < \epsilon, \ n \geq N, k = 0, 1, 2, \dots.$$

In particular,

$$|a_{Nk} - 1| < \epsilon, \ k = 0, 1, 2, \dots.$$

Thus

$$\lim_{k \to \infty} |a_{Nk} - 1| \leq \epsilon,$$

$$i.e., |0 - 1| \leq \epsilon,$$

since $A \in (\ell_\infty, c)$, $\lim_{k \to \infty} a_{nk} = 0$, $n = 0, 1, 2, \dots$, by Theorem 2 of [3],

$$i.e., 1 \leq \epsilon,$$

a contradiction on the choice of ϵ. Consequently, we have:

Theorem 105. *When K is a complete, non-trivially valued, non-Archimedean field,*

$$(\ell_1, c; P') \cap (\ell_\infty, c) = \phi.$$

Remark 23. *However, when K is a complete, non-trivially valued, non-Archimedean field,*

$$(\ell_1, c; P') \cap (c, c) \neq \phi.$$

This is amply illustrated by the following example. Consider the matrix

$$A = (a_{nk}) = \begin{bmatrix} 1 & 0 & 0 & 0 & 0 & 0 & \cdots \\ 1 & -1 & 0 & 0 & 0 & 0 & \cdots \\ 1 & 1 & -2 & 0 & 0 & 0 & \cdots \\ 1 & 1 & 1 & -3 & 0 & 0 & \cdots \\ 1 & 1 & 1 & 1 & -4 & 0 & \cdots \\ \cdots & \cdots & \cdots & \cdots & \cdots & \cdots & \cdots \end{bmatrix},$$

i.e., $a_{nk} = 1, k \leq n - 1$;

$$= -(n - 1), k = n;$$

$$= 0, otherwise.$$

Then $\sup_{n,k} |a_{nk}| \leq 1 < \infty$, $\lim_{n \to \infty} a_{nk} = 1$, $k = 0, 1, 2, \ldots$ *and* $\lim_{n \to \infty} \sum_{k=0}^{\infty} a_{nk} = 0$ *so that* $A \in (\ell_1, c; P') \cap (c, c)$, *proving the claim. Since* $(c, c) \subset (c_0, c) \subset (\ell_p, c)$, $p > 1$, *it follows that* $(\ell_1, c; P') \cap (X, c) \neq \phi$, *when* $X = c, c_0, \ell_p, p > 1$. *This indicates a violent departure in the non-Archimedean setup compared to the case* $K = \mathbb{R}$ *or* $\mathbb{C}$.

In this case, $(c_0, c; P')$ denotes the class of all matrices $A = (a_{nk}) \in (c_0, c)$ such that $\lim_{n \to \infty} (Ax)_n = \sum_{k=0}^{\infty} x_k$, $x = \{x_k\} \in c_0$. In this context, we note that $\sum_{k=0}^{\infty} x_k$ converges if and only if $\{x_k\} \in c_0$.

Remark 24. *Note that*

$$(c_0, c; P') = (\ell_1, c; P').$$

Remark 25. *We note that ℓ_p, $p \geq 1$, c_0, c, ℓ_∞ are linear spaces with respect to coordinatewise addition and scalar multiplication irrespective of the choice of K. When $K = \mathbb{R}$ or $\mathbb{C}$, c_0, c, ℓ_∞ are Banach spaces with respect to the norm*

$$\|x\| = \sup_{k \geq 0} |x_k|,$$

where $x = \{x_k\} \in c_0, c, \ell_\infty$, while they are non-Archimedean Banach spaces under the above norm when K is a complete, non-trivially valued, non-Archimedean field.

Whatever K is, ℓ_p, $p \geq 1$ is a Banach space with respect to the norm

$$\|x\| = \left(\sum_{k=0}^{\infty} |x_k|^p \right)^{\frac{1}{p}}, \quad x = \{x_k\} \in \ell_p.$$

Whatever K is, if $A = (a_{nk}) \in (\ell, c; P')$, then A is bounded and

$$\|A\| = \sup_{n,k} |a_{nk}|.$$

However, $(\ell_1, c; P')$ is not a subspace of $BL(\ell_1, c)$, i.e., the space of all bounded linear mappings of ℓ_1 into c, since $\lim_{n \to \infty} 2a_{nk} = 2$, $k = 0, 1, 2, \ldots$ and consequently $2A \notin (\ell_1, c; P')$ when $A \in (\ell_1, c; P')$.

8.4 Some more Steinhaus-type theorems over valued fields I

In this section, we introduce the sequence space

$$\gamma_\infty = \left\{ \{x_k\} : \{s_k\} \in \ell_\infty, s_k = \sum_{i=0}^{k} x_i, \ k = 0, 1, 2, \ldots \right\}.$$

We note that $\gamma_\infty \subset \ell_\infty$. $(\ell_1, \gamma; P)$ denotes the set of all infinite matrices $A = (a_{nk}) \in (\ell_1, \gamma)$ such that $\sum_{n=0}^{\infty} (Ax)_n = \sum_{k=0}^{\infty} x_k$, $x = \{x_k\} \in \ell_1$.

First, we consider the case $K = \mathbb{R}$ or $\mathbb{C}$. It is known (see [11, pp. 7, 48]) that $A = (a_{nk}) \in (\ell_1, \gamma)$ if and only if

$$\left.\begin{aligned} (i) \; & \sup_{m,k} \left| \sum_{n=0}^{m} a_{nk} \right| < \infty; \\ & and \\ (ii) \; & \sum_{n=0}^{\infty} a_{nk} \text{ converges}, \; k = 0, 1, 2, \ldots. \end{aligned}\right\} \tag{8.5}$$

We now have:

Theorem 106. *When $K = \mathbb{R}$ or $\mathbb{C}$, a matrix $A = (a_{nk}) \in (\ell_1, \gamma; P)$ if and only if it satisfies (8.5) and*

$$\sum_{n=0}^{\infty} a_{nk} = 1, \; k = 0, 1, 2, \ldots. \tag{8.6}$$

Proof. If $A \in (\ell_1, \gamma; P)$, then (8.5) (i) holds. For $k = 0, 1, 2, \ldots$, each $e_k = \{0, \ldots, 0, 1, 0, \ldots\}$, 1 occurring in the kth place, is in ℓ_1 and so $\sum_{n=0}^{\infty} (Ae_k)_n$ converges and $\sum_{n=0}^{\infty} (Ae_k)_n = 1$, $k = 0, 1, 2, \ldots$, i.e., $\sum_{n=0}^{\infty} a_{nk}$ converges and $\sum_{n=0}^{\infty} a_{nk} = 1$, $k = 0, 1, 2, \ldots$, i.e., (8.5) (ii) and (8.6) hold.

Conversely, let (8.5) and (8.6) hold. It is clear that $A \in (\ell_1, \gamma)$. Let $B = (b_{mk})$, where

$$b_{mk} = \sum_{n=0}^{m} a_{nk}, \; m, k = 0, 1, 2, \ldots.$$

Using (8.5) (i) and (8.6), we have,

$$\sup_{m,k \geq 0} |b_{mk}| < \infty$$

and

$$\lim_{m \to \infty} b_{mk} = 1, \; k = 0, 1, 2, \ldots.$$

Using Theorem 103,

$$B \in (\ell_1, c; P').$$

Let, now, $x = \{x_k\} \in \ell_1$. So

$$\lim_{m \to \infty} \sum_{k=0}^{\infty} b_{mk} x_k \text{ exists and is equal to } \sum_{k=0}^{\infty} x_k,$$

$$i.e., \quad \lim_{m \to \infty} \sum_{k=0}^{\infty} \left(\sum_{n=0}^{m} a_{nk} \right) x_k = \sum_{k=0}^{\infty} x_k,$$

$$i.e., \quad \lim_{m \to \infty} \sum_{n=0}^{m} \left(\sum_{k=0}^{\infty} a_{nk} x_k \right) = \sum_{k=0}^{\infty} x_k,$$

$$i.e., \quad \sum_{n=0}^{\infty} \left(\sum_{k=0}^{\infty} a_{nk} x_k \right) = \sum_{k=0}^{\infty} x_k,$$

$$i.e., \quad \sum_{n=0}^{\infty} (Ax)_n = \sum_{k=0}^{\infty} x_k,$$

$$i.e., \quad A \in (\ell_1, \gamma; P),$$

which completes the proof of the theorem. □

Maddox [2] proved that

$$(\gamma, \gamma; P) \cap (\gamma_\infty, \gamma) = \phi.$$

In this context, it is worthwhile to note that the identity matrix, i.e., $I = (i_{nk})$, where $i_{nk} = 1$, if $k = n$ and $i_{nk} = 0$, if $k \neq n$, is in $(\ell_1, \gamma; P) \cap (\gamma_\infty, \gamma)$ so that

$$(\ell_1, \gamma; P) \cap (\gamma_\infty, \gamma) \neq \phi.$$

Since $(\gamma_\infty, \gamma) \subset (\gamma, \gamma)$, it follows that

$$(\ell_1, \gamma; P) \cap (\gamma, \gamma) \neq \phi.$$

We have $(\gamma, \gamma; P) \subset (\ell_1, \gamma; P)$ and $(c_0, \gamma) \subset (\gamma, \gamma)$. "Enlarging" the class $(\gamma, \gamma; P)$ to $(\ell_1, \gamma; P)$, we would like to "contract" the class (γ, γ) to (c_0, γ) and attempt a Steinhaus-type theorem involving the classes $(\ell_1, \gamma; P)$ and (c_0, γ).

Theorem 107. *When $K = \mathbb{R}$ or $\mathbb{C}$,*

$$(\ell_1, \gamma; P) \cap (c_0, \gamma) = \phi.$$

Proof. Let $A = (a_{nk}) \in (\ell_1, \gamma; P) \cap (c_0, \gamma)$. Since $A \in (c_0, \gamma)$,

$$\sup_{m \geq 0} \sum_{k=0}^{\infty} \left| \sum_{n=0}^{m} a_{nk} \right| \leq M < \infty \qquad (8.7)$$

(see [11, pp. 6, 43]). Now, for $l = 0, 1, 2, \ldots, \ m = 0, 1, 2, \ldots,$

$$\sum_{k=0}^{l} \left| \sum_{n=0}^{m} a_{nk} \right| \leq \sum_{k=0}^{\infty} \left| \sum_{n=0}^{m} a_{nk} \right|$$

$$\leq M.$$

Taking the limit as $m \to \infty$, we have,

$$\sum_{k=0}^{l} \left| \sum_{n=0}^{\infty} a_{nk} \right| \leq M, \ l = 0, 1, 2, \ldots.$$

Taking the limit as $l \to \infty$, we get,

$$\sum_{k=0}^{\infty} \left| \sum_{n=0}^{\infty} a_{nk} \right| \leq M,$$

which is a contradiction, since $\sum_{n=0}^{\infty} a_{nk} = 1, \ k = 0, 1, 2, \ldots,$ in view of (8.6). This establishes our claim. $\square$

Corollary 6. *Since $c_0 \subset c \subset \ell_\infty$, $(\ell_\infty, \gamma) \subset (c, \gamma) \subset (c_0, \gamma)$ so that*

$$(\ell_1, \gamma; P) \cap (X, \gamma) = \phi,$$

where $X = c_0, c, \ell_\infty$, when $K = \mathbb{R}$ or $\mathbb{C}$.

When K is a complete, non-trivially valued, non-Archimedean field, we note that

$$\gamma = c_0 \quad \text{and} \quad \gamma_\infty = \ell_\infty.$$

In this case, it is easy to establish the following results.

Theorem 108. $(\ell_1, \gamma) = (\ell_1, c_0) = (c_0, c_0)$. *A matrix* $A = (a_{nk}) \in (\ell_1, c_0)$ *if and only if*

$$
\left.
\begin{array}{l}
(i) \ \sup_{n,k} |a_{nk}| < \infty; \\[1.5em]
and \\[1.5em]
(ii) \ \lim_{n \to \infty} a_{nk} = 0, \ k = 0, 1, 2, \dots.
\end{array}
\right\}
\tag{8.8}
$$

Theorem 109. $(\ell_1, \gamma; P) = (\ell_1, c_0; P) = (c_0, c_0; P) = (\gamma, \gamma; P)$. $A = (a_{nk}) \in (\ell_1, c_0; P)$ *if and only if (8.6) and (8.8) hold.*

Theorem 110. *A matrix* $A = (a_{nk}) \in (c, c_0)$ *if and only if (8.8) holds and*

$$
\lim_{n \to \infty} \sum_{k=0}^{\infty} a_{nk} = 0.
\tag{8.9}
$$

Theorem 107 fails to hold when K is a complete, non-trivially valued, non-Archimedean field since $(\ell_1, c_0) = (c_0, c_0)$. We also have

$$
(\ell_1, c_0; P) \cap (c, c_0) \neq \phi.
$$

This is illustrated by the following example. Consider the infinite matrix

$$
A = (a_{nk}) =
\begin{bmatrix}
1 & -1 & 0 & 0 & 0 & 0 & \cdots \\
0 & 2 & -2 & 0 & 0 & 0 & \cdots \\
0 & 0 & 3 & -3 & 0 & 0 & \cdots \\
0 & 0 & 0 & 4 & -4 & 0 & \cdots \\
\cdots & \cdots & \cdots & \cdots & \cdots & \cdots & \cdots
\end{bmatrix},
$$

$$
\begin{array}{rcll}
i.e., \ a_{nk} & = & n + 1, & \text{if } k = n; \\
& = & -(n + 1), & \text{if } k = n + 1; \\
& = & 0, & \text{otherwise.}
\end{array}
$$

Then (8.6), (8.8), (8.9) hold so that $A \in (\ell_1, c_0; P) \cap (c, c_0)$. These remarks point out a significant departure from the case $K = \mathbb{R}$ or $\mathbb{C}$.

We now prove the following Steinhaus-type result.

Theorem 111. *When K is a complete, non-trivially valued, non-Archimedean field,*

$$(\ell_1, c_0; P) \cap (\ell_\infty, c_0) = \phi.$$

Proof. Let $A = (a_{nk}) \in (\ell_1, c_0; P) \cap (\ell_\infty, c_0)$. In view of (8.6), we have,

$$1 = \left| \sum_{n=0}^\infty a_{nk} \right| \leq \sup_{n \geq 0} |a_{nk}|.$$

Taking the limit as $k \to \infty$, using (7.3), we get $1 \leq 0$, which is absurd. This proves the theorem. □

In view of Theorem 109 and Theorem 111, we have:

Corollary 7.

$$(c_0, c_0; P) \cap (\ell_\infty, c_0) = \phi.$$

8.5 Some more Steinhaus-type theorems over valued fields II

When $K = \mathbb{R}$ or $\mathbb{C}$, the following results are known (see [11, p. 4, 12 & 14]). $A = (a_{nk}) \in (c_0, c)$ if and only if

$$
\left.
\begin{array}{l}
(i)\ \sup_{n \geq 0} \left(\sum_{k=0}^\infty |a_{nk}| \right) < \infty; \\[2em]
and \\[1em]
(ii)\ \lim_{n \to \infty} a_{nk} = \delta_k \text{ exists, } k = 0, 1, 2, \ldots.
\end{array}
\right\}
\tag{8.10}
$$

$A = (a_{nk}) \in (\gamma, c, P')$ if and only if

$$\sup_{n \geq 0} \left(\sum_{k=0}^\infty |a_{nk} - a_{n,k+1}| \right) < \infty \tag{8.11}$$

and (8.10) (ii) holds with $\delta_k = 1$, $k = 0, 1, 2, \ldots$, where $(\gamma, c; P')$ is the subclass of (γ, c) with $\lim_{n \to \infty} (Ax)_n = \sum_{k=0}^\infty x_k$, $x = \{x_k\} \in \gamma$.

A Steinhaus-type result now follows.

Theorem 112.

$$(\gamma, c; P') \cap (c_0, c) = \phi.$$

Proof. Let $A = (a_{nk}) \in (\gamma, c; P') \cap (c_0, c)$. Using (8.10), it follows that

$$\sum_{k=0}^{\infty} |\delta_k| \text{ converges,}$$

which is a contradiction, since $\delta_k = 1$, $k = 0, 1, 2, \ldots$. $\square$

Corollary 8. *Since $c_0 \subset c \subset \ell_\infty$, it follows that $(\ell_\infty, c) \subset (c, c) \subset (c_0, c)$ so that*

$$(\gamma, c; P') \cap (X, c) = \phi,$$

when $X = c_0, c, \ell_\infty$.

The following results are also known (see [11, pp. 7, 47, 48]).

$A = (a_{nk}) \in (\ell_1, \gamma)$ if and only if

$$\left.\begin{array}{l} (i) \ \sup_{m,k} \left| \sum_{n=0}^{m} a_{nk} \right| < \infty; \\\\ and \\\\ (ii) \ \sum_{n=0}^{\infty} a_{nk} \text{ converges, } k = 0, 1, 2, \ldots. \end{array}\right\} \quad (8.12)$$

It is easy to prove that $A \in (\ell_1, \gamma; P)$ if and only if (8.12) holds and

$$\sum_{n=0}^{\infty} a_{nk} = 1, \ k = 0, 1, 2, \ldots, \quad (8.13)$$

where $(\ell_1, \gamma; P)$ denotes the subclass of (ℓ_1, γ) such that

$$\sum_{n=0}^{\infty} (Ax)_n = \sum_{k=0}^{\infty} x_k, \ x = \{x_k\} \in \ell_1.$$

$A = (a_{nk}) \in (\ell_p, \gamma)$, $p > 1$, if and only if (8.12) (ii) holds and

$$\sup_{m \geq 0} \sum_{k=0}^{\infty} \left| \sum_{n=0}^{m} a_{nk} \right|^q < \infty, \quad (8.14)$$

where $\frac{1}{p} + \frac{1}{q} = 1$.

Theorem 113.

$$(\ell_1, \gamma; P) \cap (\ell_p, \gamma) = \phi, \ p > 1.$$

Proof. Let $A = (a_{nk}) \in (\ell_1, \gamma; P) \cap (\ell_p, \gamma)$. Using (8.14), we can show that

$$\sum_{k=0}^{\infty} \left| \sum_{n=0}^{\infty} a_{nk} \right|^q < \infty,$$

which is a contradiction, since $\sum_{n=0}^{\infty} a_{nk} = 1$, $k = 0, 1, 2, \ldots$, using (8.13). This completes the proof. $\square$

Corollary 9. *Since $\ell_p \subset c_0 \subset c \subset \ell_\infty$, it follows that $(\ell_\infty, \gamma) \subset (c, \gamma) \subset (c_0, \gamma) \subset (\ell_p, \gamma)$ so that*

$$(\ell_1, \gamma; P) \cap (X, \gamma) = \phi,$$

when $X = \ell_p, c_0, c, \ell_\infty$.

Again, it is known (see [11, pp. 6, 43]) that $A = (a_{nk}) \in (c_0, \gamma)$ if and only if (8.12) (ii) holds and

$$\sup_{m \geq 0} \sum_{k=0}^{\infty} \left| \sum_{n=0}^{m} a_{nk} \right| < \infty. \tag{8.15}$$

It is also known that $A = (a_{nk}) \in (\gamma, \gamma)$ (see [11, pp. 7, 45]) if and only if (8.12) (ii) holds and

$$\sup_{m \geq 0} \sum_{k=0}^{\infty} \left| \sum_{n=0}^{m} (a_{nk} - a_{n,k-1}) \right| < \infty. \tag{8.16}$$

We can prove that $A \in (\gamma, \gamma; P)$ if and only if (8.13) and (8.16) hold, where $(\gamma, \gamma; P)$ is the subclass of (γ, γ) such that $\sum_{n=0}^{\infty} (Ax)_n = \sum_{k=0}^{\infty} x_k$, $x = \{x_k\} \in \gamma$.

Consequently, we have the following Steinhaus-type result.

Theorem 114.

$$(\gamma, \gamma; P) \cap (c_0, \gamma) = \phi.$$

Proof. Let $A = (a_{nk}) \in (\gamma, \gamma; P) \cap (c_0, \gamma)$. Using (8.15), it follows that

$$\sum_{k=0}^{\infty} \left| \sum_{n=0}^{\infty} a_{nk} \right| < \infty,$$

which is a contradiction, since $\sum_{n=0}^{\infty} a_{nk} = 1$, $k = 0, 1, 2, \ldots$, in view of (8.13).

$\square$

Corollary 10. *We have*

$$(\gamma, \gamma; P) \cap (X, \gamma) = \phi,$$

when $X = c_0, c, \ell_\infty$.

$A = (a_{nk}) \in (c_0, \ell_1)$ (see [11, pp. 8, 72]) if and only if

$$\sup_{N} \sum_{k=0}^{\infty} \left| \sum_{n \in N} a_{nk} \right| < \infty, \tag{8.17}$$

where N is a subset of $\mathbb{N}_0 = \{0, 1, 2, \ldots\}$.

$A = (a_{nk}) \in (\gamma, \ell_1)$ (see [11, pp. 9, 74]) if and only if

$$\sup_{N, K} \left| \sum_{n \in N} \sum_{k \in K} (a_{nk} - a_{n,k-1}) \right| < \infty, \tag{8.18}$$

where N, K are subsets of $\mathbb{N}_0$. We can prove that $A \in (\gamma, \ell_1; P)$ if and only if (8.13) and (8.18) hold, where $(\gamma, \ell_1; P)$ is the subclass of (γ, ℓ_1) such that $\sum_{n=0}^{\infty} (Ax)_n = \sum_{k=0}^{\infty} x_k$, $x = \{x_k\} \in \gamma$.

Theorem 115.

$$(\gamma, \ell_1; P) \cap (c_0, \ell_1) = \phi.$$

Proof. If $A = (a_{nk}) \in (\gamma, \ell_1; P) \cap (c_0, \ell_1)$, using (8.17), we have,

$$\sum_{k=0}^{\infty} \left| \sum_{n=0}^{\infty} a_{nk} \right| < \infty,$$

which is a contradiction, in view of (8.13).

$\square$

Corollary 11.

$$(\gamma, \ell_1; P) \cap (X, \ell_1) = \phi,$$

when $X = c_0, c, \ell_\infty$.

Remark 26. $(c, \gamma; P')$ *denotes the subclass of* (c, γ) *with* $\sum_{n=0}^{\infty} (Ax)_n = \lim_{k \to \infty} x_k$, $x = \{x_k\} \in c$. *In the context of Steinhaus-type theorems, we note that*

$$(c, \gamma; P') \cap (\ell_\infty, \gamma) \neq \phi.$$

In the rest of this section, K denotes a complete, non-trivially valued, non-Archimedean field. The following observations are worth recording in the context of Theorem 112.

In this case, $\gamma = c_0$ and so

$$(c_0, c; P') \cap (c_0, c) = (c_0, c; P') \neq \phi.$$

We can prove that

$$(c_0, c; P') \cap (c, c) \neq \phi,$$

for which we need the following example: Consider the infinite matrix

$$A \equiv (a_{nk}) = \begin{bmatrix} 1 & -1 & 0 & 0 & 0 & 0 & \cdots \\ 1 & 1 & -2 & 0 & 0 & 0 & \cdots \\ 1 & 1 & 1 & -3 & 0 & 0 & \cdots \\ 1 & 1 & 1 & 1 & -4 & 0 & \cdots \\ \cdots & \cdots & \cdots & \cdots & \cdots & \cdots & \cdots \end{bmatrix}.$$

It is clear that $\sup_{n,k} |a_{nk}| < \infty$, $\lim_{n \to \infty} a_{nk} = 1$, $k = 0, 1, 2, \ldots$ and $\lim_{n \to \infty} \sum_{k=0}^{\infty} a_{nk} = 0$. Thus $A \in (c_0, c; P') \cap (c, c)$ (see [4]), proving the claim. However, the following Steinhaus-type result holds in the non-Archimedean setup.

Theorem 116. *When* K *is a complete, non-trivially valued, non-Archimedean field,*

$$(c_0, c; P') \cap (\ell_\infty, c) = \phi.$$

Proof. Let $A = (a_{nk}) \in (c_0, c; P') \cap (\ell_\infty, c)$. Since $A \in (\ell_\infty, c)$, $\sum_{k=0}^{\infty} \delta_k$ converges, where $\lim_{n \to \infty} a_{nk} = \delta_k$, $k = 0, 1, 2, \ldots$. This is a contradiction, since $\delta_k = 1$, $k = 0, 1, 2, \ldots$, completing the proof.   □

In view of Remark 3.4 of [6],

$$(\ell_1, c_0: P) \cap (X, c_0) \neq \phi,$$

when $X = \ell_p, c_0, c$, $p > 1$, while

$$(\ell_1, c_0; P) \cap (\ell_\infty, c_0) = \phi$$

(see [6, Theorem 3.6]).

Noting that $(c_0, c_0; P) = (\ell_1, c_0; P)$, we can write the above results with $(\ell_1, c_0; P)$ replaced by $(c_0, c_0; P)$.

We note that

$$(c_0, \ell_1; P) \cap (c_0, \ell_1) = (c_0, \ell_1; P) \neq \phi$$

(compare with Theorem 115). However, we have the following result.

Theorem 117.
$$(c_0, \ell_1; P) \cap (c, \ell_1) = \phi.$$

Proof. Let $A = (a_{nk}) \in (c_0, \ell_1; P) \cap (c, \ell_1)$. Since $A \in (c_0, \ell_1; P)$, (8.13) holds. Since $A \in (c, \ell_1)$,

$$\sum_{n=0}^{\infty} \sum_{k=0}^{\infty} a_{nk} \quad \text{converges.}$$

In view of the fact that convergence is equivalent to unconditional convergence (see [12, p. 133]),

$$\sum_{k=0}^{\infty} \sum_{n=0}^{\infty} a_{nk} \quad \text{converges,}$$

which is a contradiction, since, by (8.13), $\sum_{n=0}^{\infty} a_{nk} = 1$, $k = 0, 1, 2, \ldots$, completing the proof.   □

Corollary 12. *Since $c \subset \ell_\infty$, $(\ell_\infty, \ell_1) \subset (c, \ell_1)$ so that*

$$(c_0, \ell_1; P) \cap (X, \ell_1) = \phi,$$

when $X = c, \ell_\infty$.

Remark 27. *Analogous to the class $(c, \gamma; P')$, when $K = \mathbb{R}$ or $\mathbb{C}$, we may be tempted to consider the class $(c, c_0; P')$, when K is a complete, non-trivially valued, non-Archimedean field. However, we note that, in this case,*

$$(c, c_0; P') = \phi.$$

For, if $A = (a_{nk}) \in (c, c_0; P')$, we have,

$$\sum_{n=0}^{\infty} a_{nk} = 0, \ k = 0, 1, 2, \ldots \tag{8.19}$$

and

$$\sum_{n=0}^{\infty} \sum_{k=0}^{\infty} a_{nk} = 1. \tag{8.20}$$

Now,

$$1 = \sum_{n=0}^{\infty} \sum_{k=0}^{\infty} a_{nk}, \quad using \ (8.20)$$

$$= \sum_{k=0}^{\infty} \sum_{n=0}^{\infty} a_{nk}, \quad since \ convergence \ is \ equivalent \ to$$

$$unconditional \ convergence \ (see [12, \ p. \ 133])$$

$$= 0, \quad using \ (8.19),$$

which is absurd, proving the claim.

In conclusion, the author wishes to point out that for a study of double sequences and double series, in the context of a new definition of convergence of a double sequence, one can refer to [9] for the classical case and [8] for the non-Archimedean case. For a detailed study of special summability methods in complete, non-trivially valued, non-Archimedean fields, one can refer to [8].

Bibliography

[1] J.A. Fridy. Properties of absolute summability matrices. *Proc. Amer. Math. Soc.*, 24:583–585, 1970.

[2] I.J. Maddox. On theorems of Steinhaus type. *J. London Math. Soc.*, 42:239–244, 1967.

[3] P.N. Natarajan. The Steinhaus theorem for Toeplitz matrices in non-archimedean fields. *Comment. Math. Prace Mat.*, 20:417–422, 1978.

[4] P.N. Natarajan. Some Steinhaus type theorems over valued fields. *Ann. Math. Blaise Pascal*, 3:183–188, 1996.

[5] P.N. Natarajan. A theorem of Steinhaus type. *J. Analysis*, 5:139–143, 1997.

[6] P.N. Natarajan. Some more Steinhaus type theorems over valued fields *Ann. Math. Blaise Pascal*, 6:47–54, 1999.

[7] P.N. Natarajan. Some more Steinhaus type theorems over valued fields II. *Comm. Math. Analysis*, 5:1–7, 2008.

[8] P.N. Natarajan. An introduction to ultrametric summability theory. Second Edition, Springer, 2015.

[9] P.N. Natarajan. *Classical summability theory*. Springer, 2017.

[10] P.N. Natarajan. Steinhaus type theorems over valued fields: A survey. *Advanced Topics in Mathematical Analysis*, ed. by M. Ruzhansky and H. Dutta, pp. 529–562, Taylor and Francis, 2018.

[11] M. Stieglitz and H. Tietz. Matrixtransformationen von Folgenräumen Eine Ergebmisübersicht. *Math. Z.*, 154:1–16, 1977.

[12] A.C.M. van Rooij and W.H. Schikhof. Non-archimedean analysis. *Nieuw Arch. Wisk.*, 19:120–160, 1971.

Index